BECOMING AN EFFECTIVE RIDER

RIDER

Developing Your Mind and Body for Balance and Unity

By Cherry Hill

Illustrations by Ann Blackstone
Photographs by Cherry Hill and Richard Klimesh

A GARDEN WAY PUBLISHING BOOK

STOREY

STOREY COMMUNICATIONS, INC.
POWNAL, VT 05261

The mission of Storey Communications is to serve our customers
by publishing practical information that encourages personal independence
in harmony with the environment.

Cover design by Meredith Maker
Front cover photographs of Carla Wennberg (upper left) and Kelly McElwie (upper right) by Cherry Hill; Kathleen Donnelly (lower left) and Cherry Hill (lower right) by Richard Klimesh
Back cover photographs by Cherry Hill (two upper) and Richard Klimesh (bottom)
Text design and production by Michelle Arabia
Edited by Deborah Burns
Interior photographs by Cherry Hill and Richard Klimesh
Interior illustrations by Ann Blackstone
Indexed by Joyce Goldstern

The information in this book is true and complete to the best of our knowledge. All recommendations are made without guarantee on the part of the author or Storey Communications, Inc. The author and publisher disclaim any liability with the use of this information. For additional information please contact Storey Communications, Inc., Schoolhouse Road, Pownal, Vermont 05261.

Where brand names appear in this book and Storey Communications, Inc. is aware of a trademark claim, those names have been printed in initial caps.

Printed in the United States by Capital City Press
Eighth Printing, June 1995

Garden Way Publishing was founded in 1973 as part of the Garden Way Incorporated Group of Companies, dedicated to bringing gardening information and equipment to as many people as possible. Today the name "Garden Way Publishing" is licensed to Storey Communications, Inc., in Pownal, Vermont. For a complete list of Garden Way Publishing titles call 1-800-827-8673. Garden Way Incorporated manufactures products in Troy, New York, under the TROY-BILT® brand including garden tillers, chipper/shredders, mulching mowers, sicklebar mowers, and tractors. For information on any Garden Way Incorporated product, please call 1-800-345-4454.

Library of Congress Cataloging-in-Publication Data

Hill, Cherry, 1947-
    Becoming an effective rider: developing your mind and body for balance and unity / by Cherry Hill : illustrations by Ann Blackstone : photographs by Cherry Hill and Richard Klimesh.
        p.   cm.
    Includes bibliographical references (p.   ) and index.
    ISBN 0-88266-689-4 (hc) — ISBN 0-88266-688-6 (pbk.)
    1. Horsemanship.   I. Title.
SF309.H63   1991
   798.2'3—dc20   91-55078
                                                                       CIP

CONTENTS

DEDICATION

To Zinger, Sassy, and Zipper.
And to dedicated riders everywhere.

ACKNOWLEDGMENTS

Thank you to the following people for their help in the preparation of this manuscript:

Richard Klimesh
RaeAnn Curtis
Carla Wennberg
Barby Fairbanks Eide

Thank you to the following photograph models:

Richard Klimesh
Carla Wennberg
Margot McAllister
RaeAnn Curtis
Sue Dixon
Barby Fairbanks Eide
Leroy Michael Eide
Kathleen Donnelly
Kevilina Burbank
Tyffani Hansen
Holly Hansen
Michael Lieberg
Kelly McElwie

PREFACE

As with many people, my earliest riding consisted of bareback and western riding through pastures, next to roads, and along trails. When I wanted to refine my skills and participate in horse shows, I rode western horsemanship, western pleasure, and trail, and then hunt seat equitation and hunt seat pleasure. To get a feel for more specialized forms of riding, I tried working cow horse, reining, and eventing. These experiences have given me an invaluable understanding and appreciation for many types of horse performances and have made me somewhat of an eclectic equestrienne.

After being a trainer, instructor, and judge for more than fifteen years, I began writing about horses. Lately I have realized that my personal involvement with riding has come full circle. I began as a recreational rider and have become a recreational rider once again. I ride now out of choice, not out of necessity. I find riding a wonderful diversion in a tightly packed schedule of writing deadlines, lec-

ture and seminar preparations, and judging assignments. Although I ride almost every day and approach it in a disciplined manner, riding is much more enjoyable for me today than it has ever been. Presently I am interested in non-competitive dressage and mountain riding.

Although I am intrigued by and in awe of top quality riders in many events, I design my riding program to suit my own interests, goals, facilities, climate, and schedule. Mountain riding provides me with a chance to use my horses in the "real world." Although mountain rides are often scenic and relaxing, the sudden appearance of a rattlesnake, rock slide, or hailstorm can create exciting and perhaps even dangerous situations. In a very real way, trail riding can serve as a test of a horse's training and a rider's ability.

Because of my teaching and training background, I can appreciate regular arena work as part of my riding program. A few years ago, like many other American riders, I took a closer look at dressage as a system for training horses. The more I learned about dressage in its classic sense, the more intrigued I became by its discipline and effectiveness. Especially during my first year of dressage lessons, I became humbly aware of the many aspects of my riding and training that needed improvement. As has often been said of many life pursuits, the more you know, the more you realize how much there is to know. Well, in my case, it seemed I knew the principles but had difficulty getting my body and my horses to put them into practice!

The changes that I needed to make included altering or eliminating habits that I had acquired during previous styles of riding. I found it necessary to retrain my horses in order to alter their western-style responses so they would react more suitably for dressage. In many instances, that meant starting from square one. In some ways I think relearning is more difficult than initial learning for both horse and rider because previously ingrained motor skills can be resistant to change. A similar type of relearning would be required if a cutting horse rider decided to pursue jumping, or an endurance racer took a sudden interest in show ring pleasure horses. Switching between styles of riding may require hard work but it can also create new avenues for learning. Being a student once again provided me with valuable insights into the trials and tribulations as well as the successes and elations of learning how to ride.

Among and within the different types of riding, there is a great variation in tack, training techniques, and rider influences. It would be impossible to cover such information in total and it would be inappropriate to emphasize horse training in a book on the development of the rider. So, rather than state the specific position in which the rider's hands or ankles should be (as these vary greatly from one style of riding to another) I am going to concentrate on the mental and physical development of the rider as a student and an athlete.

You do not need to be an aspiring professional or a full-time rider to be a dedicated, serious rider. In fact, one of the reasons I feel compelled to write this book for part-time recreational riders is because that is what most of us are. Since I spend the majority of my time as a writer, judge, and clinician, I am a part-time rider. Therefore, I probably experience time limitations, schedule crunches, and family demands similar to yours.

The novice and the accomplished professional have a lot more in common, however, than might initially be perceived. All riders experience similar feelings of exhilaration when they reach their goals. And most riders have experienced the frustration of a seemingly unsolvable problem or a plateau. No matter what the level of your riding skills, you are entitled (and I encourage you) to be excited when you make progress. Conversely, when things are not working well for you, it is OK to feel discouraged — frustration can serve as powerful fuel for motivation.

Becoming an effective rider involves intellectual, physical, moral, and emotional development. It has the potential to enhance all aspects of life. I feel very strongly that riding has helped me to become a better person. When I approach riding with dedication, I see positive changes in myself; I have seen similar changes take place in other riders of all ages and skill levels.

In our highly technical society, we commonly value the read-out from machines and instruments more highly than information from other people, animals, and our own instincts. Contrary to what people may jokingly say, it is unfortunate that we have lost many of our animal instincts. The phrase "animal instincts" often conjures up a very negative image — that of a barbarian with crude urges and actions. In fact, a person who has retained animal instincts is often more *sensitive* — more aware of the subtleties of sight, hearing, feeling, taste, and smell. This person has not blocked off the intuitive voice with supposed absolute facts. Animal instincts can guide and guard us and help distil our lives into clearer, simpler terms. And nowhere can we be better reminded of our inherent animal qualities and expressions than when riding horses.

Riding, however, is only one part of the horse/human equation. Equally important to the formula for a top horseman is knowledge and experience in horse behavior and horse care. There is a great deal of important information about psychology, tack, and training that can transform a person from a mere passenger to an effective rider. Similarly, there are specialized facts concerning nutrition, health, and management of horses that are essential for proper horse care. Because BECOMING AN EFFECTIVE RIDER focuses very specifically on the mental and physical development of the rider, I have not included horse training and management information in this book. But I strongly emphasize that in order to become a top-notch rider, you must become a conscientious all-around horse person. I encourage you to read the training and management books that I have listed in the bibliography and develop the horse management skills necessary to care for your horse properly.

My ideas are based on fifteen years of observing hundreds of my own students as well as the students of my colleagues. In addition, I continue to learn each day as I ride. To learn is to be alive.

INTRODUCTION

Humans have been riding horses for more than 4,000 years. Historically, horses were used out of necessity: for transportation, war, and hunting. Today riding is mainly recreational, although the intensity and the style of the pursuit varies widely among hobbyists. Your reasons for riding will determine your approach to it. Are you a recreational, competitive, or artistic rider? These are not mutually exclusive categories; many riders see themselves having characteristics from all three categories.

A recreational rider often looks at riding as an occasional, leisure-time activity of relatively low physical and mental intensity. The primary goal of the recreational rider is relaxation and enjoyment of a pleasurable, non-stress inducing activity.

The competitive rider is typically very goal-oriented and works regularly toward each goal with both mental and physical intensity. A realist, the competitive rider is often very focused with the end clearly in sight. Unfortunately, as with other goal-oriented activities, the competitive rider can put too much emphasis on the goal itself and forget the importance of the process of getting there. If not kept in perspective, competition can be a limiting factor in the development of a rider. Approached properly, it can be an aid to achievement and a way to exemplify and develop good sportsmanship.

The artistic rider is most interested in aesthetics: beauty, form, and harmony between horse and rider. The artistic rider is often an idealist, approaching riding with a deep sense of dedication. Although an artistic rider's involvement is often very disciplined and intense, when the partnership between the rider and the horse attains true harmony, the performance appears effortless. A truly artistic rider performs gracefully in a lyrical sense.

Using terms such as pleasure rider, dressage rider, show rider, backyard horseman, and trail rider to classify a rider can be misleading. These words have too many meanings to accurately describe the level of involvement of a rider. For example, a person who rides an unclipped, unshod, quiet farm horse across a pasture is termed a pleasure rider. Yet a person who rides a highly trained, meticulous, shiny, show-ring horse that performs lightly, fluidly, and in balance at all gaits is also termed a pleasure rider, although the two could hardly be more different.

A person who "rides dressage" could be someone who uses basic dressage principles in a broad sense in the training of any horse, or it could refer to someone who is practicing dressage in its most highly refined, specialized European-based sense. A trail rider could be a casual meanderer or an intense endurance ride competitor. A backyard horseman could be a young child with her first pony or a life-long accomplished horseman with a small acreage.

NO ABSOLUTES

With riding and horse training, there are no absolute rules. What may work wonders with one horse and one rider in one situation may be completely wrong in another set of circumstances. Learning how to ride a horse "on the bit," for example, is excellent for dressage lessons in a ring, but the same principles would be difficult and inappropriate to use when shinnying down a rock slope.

Similarly, the recipe-like instructions that are included in most how-to books on riding are based on ideal situations and must be modified each time they are applied. For example, at the very beginning of things, the instructions for mounting

may direct the rider to keep light contact with the left hand on the reins. But if the horse backs up as the rider begins to mount, then the rider must loosen the contact on the reins, perhaps to the point of slack reins if that is how the horse has been trained. The rider must adapt ideal rules to each situation in order to "get the job done" — in this case, to mount. The ideal provides a rider with a safe, tested, recommended method — a goal to work toward — but sometimes the ideal is not appropriate if followed to the letter. Ideal principles must be interpreted and adjusted according to each rider's physique, skill level, mental attitude, the horse's physique and level of training, and the tack being used.

When comparing western, hunt seat, saddle seat, and dressage styles of riding, the NO ABSOLUTES rule is particularly obvious. Since there is much variation among riding styles regarding training techniques, tack, rider position, and intended purpose, it would be futile to try to set an absolute standard for the proper length of the stirrups, for example. A dressage-length stirrup might be dangerously long for jumping and an endurance-length stirrup would be inappropriate for cutting. Additionally, for security a beginner rider in any discipline often rides with shorter stirrups than a more advanced rider.

In spite of the wide variation in riding styles, some basic concepts of rider development and horse handling are common to all forms of riding. Your goal might be to develop your own personal and effective riding style. Universal concepts and principles will be applicable whether you are a beginning or advanced rider.

Beginner, Intermediate, or Advanced?

One of the first steps in developing any skill is to determine where you are starting. Although the words beginner, intermediate, and advanced are vague, see if the descriptions that follow help you to find where you presently fit. Persons of all ages can be found in any of the categories. Some people feel that any person who is riding a horse is also training that horse — whether purposely or inadvertently and whether good or bad habits are being formed. Theoretically, I agree with that. However, I use the word trainer for those riders who have advanced enough in their skills that they can ride a variety of horses well and that they have a very good chance of eliciting the desired response from a horse the first or second time they ask the horse to perform a specific maneuver.

The PRE-BEGINNER rider is someone who is interested and curious, but totally inexperienced with horses and needs to learn about them from the ground up. She is learning how to lead horses, groom them, and relate to their size and movement from the ground. The pre-beginner rider has no knowledge about horse training and care. She may be timid or fearless.

The BEGINNER rider is entering the awareness-development stage. She might have spent a total of 10 hours in the saddle, either by riding once in a while over a period of years or by taking an introductory group of lessons. She can control a quiet school horse at a walk and trot with turns and can stop the horse. The beginner rider slows down and walks if she feels she is losing her balance at the trot. She will remain a beginner until she has developed the balance and confidence to lope or canter the horse.

The ADVANCED BEGINNER can sit on a quiet, well-schooled horse without losing her balance when it is loping or cantering.

The INTERMEDIATE rider may show signs of competitiveness or seriousness about riding. When she works with a school horse, she can mount without assistance, walk, trot, canter, ride circles, serpentines, knows what diagonal or lead she is on, and can stop the horse from any gait.

The ADVANCED INTERMEDIATE rider can perform simple and flying lead changes, variations in the gaits, turn on the forehand, turn on the hindquarters, and lateral movements on a school horse. She has the interest to ride horses other than school horses and she is gaining the knowledge and is developing the skills to be able to train horses.

The ADVANCED rider is a horse trainer with a well-developed sense of balance and timing. She understands sophisticated concepts of horse movement and sound principles of training and can perform the advanced maneuvers in her style of riding. When she rides it is very difficult to see the aids she is giving her horse.

The Serious, Dedicated Rider

Do you want to be a serious, dedicated rider? A rider with any degree of experience can become one. Do not think that becoming serious or dedicated about your riding will mean that you will lose your sense of humor or that you will be required to live, breathe, and talk about nothing but horses. Quite the contrary. Some of the most successful riders say that a sense of humor is what helps them to keep the ups and downs of their work in perspective. And as far as having interests other than riding, it is essential! Only by being a well-rounded individual will you bring to your riding an overall sense of well-being. A serious, dedicated rider is a person who has chosen riding as a recreational vocation — an activity pursued for self-development and self-satisfaction as well as relaxation and enjoyment.

The dedicated rider is a life-long learner, intent on maximizing her equine knowledge and skills. A dedicated beginning rider is not a lesser person than a more skilled rider. Dedicated riders at all levels have certain characteristics in common: a healthy self-image, a consistently positive attitude toward work, and a knowledge of successful principles for dealing with themselves, their horses, and other people.

The sound physical and mental self-image of the dedicated rider makes daily tasks run smoothly and adds a measure of help when problems arise. Problems are part of learning how to ride. How you react to a problem will greatly affect the future of your endeavor. The dedicated rider is not afraid of change, as it often leads to growth and improvement. Individuals who view problems as opportunities for learning rather than deterrents are ultimately more successful. People that like things to stay the same tend to progress more slowly. Remember, when the going seems easy, it may just be that you are going downhill.

Evaluations are an essential part of becoming a better rider. Learning how to appraise your own skills will be discussed later in the book. A developing rider needs more, however, than self-evaluation. Critiques from qualified instructors are essential. Riders at all levels have room for improvement and should receive warranted criticism with respect and an open mind. During an evaluation, apologies are unnecessary and excuses are non-productive. Instead, focus your energies specifically on what you can do to improve.

Often the needed changes involve habits that occur out of the saddle. The rider's body is most effective when maintained by moderation and regularity in eating, drinking, sleeping, and exercising. Maximum performance is contingent on dedication to healthy habits. A healthy physical self-image begins with high standards of personal hygiene and a tidy appearance. Although what's on the inside counts, what's on the outside shows. If you are sloppy or careless in your dress, it can cause you to approach your work with horses in the same manner. Additionally, if your personal appearance is offensive you may alienate your fellow students or your instructor.

The dedicated rider really enjoys her involvement with horses. Horses provide a good way for you to get to know yourself and they can offer a way for you to reach some of your personal goals. The successful rider at any level knows the answers to the following questions: Where have I been? Where am I now? Where do I want to go? How do I get there?

GOALS

The process of reaching goals includes an initial evaluation, frequent reviews, and progress checks. Goals should be set down in specific terms so they appear as crystallized pictures in the mind rather than fuzzy apparitions on the horizon. "I have to become a better rider" sounds like a project of enormous proportions with nowhere to start. Setting a more concrete, short-term goal is more effective. For example, decide that at the end of two weeks you will be able to effectively ride a horse 5 strides canter, 3 strides trot, 5 strides canter, 3 strides trot, etc. for one entire round of the arena. This is a more specific, practical, and therefore attainable goal.

Obstacles should be viewed realistically. It is a common error to visualize small hurdles as massive stone walls. This is especially true of riders who have had a bad experience with a horse. If you worry that a horse is going to bite your finger (whether purposefully or accidentally) when you are bridling him, and this puts you in a nervous state, take one or two lessons that emphasize practicing bridling with and without assistance so that you can confidently tack your horse for future rides. Rather than fabricate an insurmountable wall in your mind which hinders you in all future lessons, step up to a problem, see what it requires, and formulate a plan.

When a large or lifetime goal is broken up into smaller pieces so that it is attainable in stages, it is more likely to be reached. Whether you reach your goal is largely dependent on your ability to manage your assets and resources. Most important is how you "spend" your time. Although you may not have control of how inflation affects the cost of hay and a saddle, you still govern your personal balance in your daily bank of time. Successful time management depends on your ability to focus on effectiveness rather than merely efficiency. Efficiency is doing something well. Effectiveness is doing the right thing well.

The dedicated rider knows how to target time each day toward important, high-pay-off activities rather than getting caught up in a false sense of duty and a long list of less important activities. It can be easy to find yourself spending a disproportionate amount of time cleaning tack, washing blankets and bandages, grooming, and talking about riding when it is actual riding that you should be concentrating on.

Alternatively, it might be easy for you to get caught up in an intense enthusiasm for riding and want to "get there" fast. Although there are many shortcuts used to attain temporary and superficial successes, there is no substitute for time in the saddle to develop a rider. Sometimes if you get going too fast, you will misplace your focus, lose sight of where you were initially headed, and get derailed on a non-productive side track. For example, a dressage rider who practices only shoulder-in-left before a dressage test may find that once in the ring her horse has difficulty going straight. This rider has lost her overall perspective. A mental rehearsal at the beginning of each day and before each ride will help you to identify your priorities and will result in good use of your time. Mental planning is especially critical immediately preceding lessons or competitions.

Organization and prioritizing are imperative to the success of the peak performer. Twenty percent of the things you do yield eighty percent of your gain.

Attending to the most important items on the barn "TO DO" list will result in maximum productivity. Less essential items can be ticked off as time permits. Plan the work and then work the plan.

Why is it that women riders outnumber men ten to one? Part of it is probably due to the nurturing, caring relationship that can develop between a horse and its rider. But part of it may be a way for women to further dilute the differences society has assigned to the sexes. Sports, and riding in particular, may appeal to women because they allow women to step out of two female stereotypes — passivity and dependence. Taking charge of a 1,200-pound animal can hardly be considered passive or dependent.

FACTORS IMPORTANT TO THE SUCCESS OF THE RIDER

Over the years, I have observed that most riders face common obstacles. However, rather than presenting my observations to you as a series of problems that you must learn to overcome, I'm listing them as factors that are important to the successful development of a rider. After each factor are descriptions of people at the opposite ends of the spectrum regarding that point. These factors are discussed elsewhere in the book in more detail.

A positive, confident attitude. The person who faces the jump expecting to land correctly on the other side, vs. the person who remembers the time her horse refused or went over the jump badly.

A healthy ego. The person who has pride in her work and works to satisfy her own standards, vs. the show-off or know-it-all whose primary reason for performing is to show others her supposed superiority.

A fit and healthy body. The person who can breathe and cool her body during exercise and who is supple and of a healthy weight, vs. the person who overheats or is short of breath, stiff, or overweight.

Open-mindedness. The person who accepts advice as a gift, vs. the person who interprets advice as an insult or threat.

A pleasant yet strong personality. The person who says hello, is willing to help, yet goes about her work, vs. the person who is snooty, ignores most people, yet wastes time in idle gossip with a few.

Self-motivation. The person who works after school to earn lessons, vs. the person who must be pushed and prodded by parents or friends to participate.

Responsibility for hard work. The person who cleans the stall and picks out the horse's hooves each day, vs. the person who does it only when told or when the horse is sick or lame.

Emotional security. The person who reacts to disappointment with a plan to improve, vs. the person who reacts to disappointment with crying or losing her temper.

Mental alertness. The person who understands what the instructor is asking her to do, vs. the person who repeatedly asks the instructor to explain the same things and cannot follow instructions.

The ability to analyze and self-evaluate. The person who can work at home and assess the quality of the work during and after the session, vs. the person who is afraid to try anything without someone there to tell her if she is performing it correctly.

The ability to plan a work session. The person who is able to choose a productive plan: the warm-up, the old work, the new work, and the cool-down, vs. the person who rides aimlessly into an arena and wonders, "Now what do I do?"

Interest in horse behavior, training principles, tack. The person who spends time reading, observing, and talking with qualified horse people, vs. the person who gets by with just what is needed for basic participation.

Time. The person who plans regular, unhurried blocks of time, vs. the person who makes time once in a while or uses short periods of time sandwiched hectically between other activities.

Adequate finances. The person who uses the money available to care properly for a horse and participate in an activity at an appropriate level, vs. the person who uses money for less important, superficial purchases while the level of the horse's care suffers.

A good school horse. A cooperative, well-trained horse of any breed or appearance, vs. a handsome or beautiful but flighty and insufficiently or improperly trained horse.

A good instructor. A qualified person able both to tell and show you how to ride, vs. a questionable, domineering individual who has difficulty explaining but can ride, or is a great talker but can't ride well.

A place to work on a regular basis. A level, uncongested, enclosed area with good footing that is always available, vs. a rough or wet area with unsafe fencing and inconsistent availability.

The proper tack. A saddle of the proper size and style that lets you sit deep in the middle of the seat and allows some degree of flexibility, vs. a saddle that is too big or too small or locks you into an incorrect, fixed position.

DEVELOPING YOUR PERSONALIZED APPROACH TO RIDING

In order to be most effective, the programs of mental and physical development that you choose to follow should be based on your personal philosophies and goals. This will give you the greatest chance for success. It would make no sense to choose a mental approach based on Zen or yoga if you feel uncomfortable just hearing words such as "centering" or "energy body." However, if an American style of mental training consisting of goal-setting and time management feels more like a business meeting to you, then that might not be an appropriate pattern to follow either. You may have to combine various ideas and customize them so that you are comfortable with the plan. That way you will use it.

Similarly, when it comes to choosing your physical development plan, you will have to do some customizing. Your goals can be reached using a combination of isometrics, isotonics, aerobics, yoga, walking, cross-training with a companion sport, and riding. The important thing is to design a physical plan that you will be most likely to follow on a regular basis.

There are many approaches to sports and to riding in particular, and because of each rider's experience level, everyone has different needs. Rather than suggesting one plan over another, I'll discuss some of the major concepts and techniques surrounding the mental and physical training of athletes and relate them specifically to riding. That way, you can select and blend your own personalized riding formula. I have found that a blend of techniques and approaches works best for most students, including myself.

I will use selected how-to examples to explain the principles that are involved in learning how to ride. As you read, I encourage you to reach out for those ideas that are the most unfamiliar to you. Read with an open mind. By viewing new information as a gift for your growth rather than as a threat to your comfort, you may discover an important key to your riding that has previously eluded you.

SECTION ONE

MENTAL DEVELOPMENT

CHAPTER 1

MENTAL TRAITS

Certain mental characteristics will allow a person to be more successful as a rider and, not surprisingly, riding will serve to develop those traits more fully in a person. Research has shown that the pursuit of athletic activities makes a person more confident, sociable, and extroverted than a non-athletic person, better at problem solving and self-evaluation. Athletes are generally less neurotic and more emotionally stable than non-athletes. But if an athletic pursuit such as riding is carried to an extreme, it can create harmful stresses. The excessive physical and mental exertion of fanaticism can result in physical and emotional problems. That is why it is best to work toward integrating riding into the rest of your life. Balancing your professional, recreational, personal, and family endeavors gives you a better chance of being a well-rounded, happy, and healthy person. The following mental and emotional traits are of great importance to the developing rider.

Motivated. Motivation is the desire, the "want to," the inspiration behind the actions, the incentive to participate and to improve. There can be several different, legitimate reasons behind any person's motivation for becoming a rider. It can simply be the love of horses — the feeling one experiences being near them. It can be admiration of a particular equestrian sport. It can be the urge to be physically active. It can be a desire for social approval, success, or prestige. It may be a way to improve self-confidence and self-respect. Or it may be to attract attention from the opposite sex. Riding can represent an escape from the pressures of a job, a marriage or relationship, or adolescence. Or it may provide a diversion from a seemingly boring existence. A handful of people pursue riding in the hope of receiving monetary rewards for their efforts. Whatever the reason for the initial interest in riding, if a person pursues it with energy and interest, riding can give much in return.

The enthusiastic person generates an aura of sincere interest and an eager desire to learn. The keen rider is constantly

Positive attitude.

looking for opportunities to discover new things about the way her mind and body affect a horse's attitude and way of moving. A motivated rider is a "self-starter," someone who receives inspiration from within and does not rely on someone or something else to provide the incentive to continue the study of riding. An occasional boost from an instructor or a friend is fine, but if you require their daily encouragement to keep up your interest, you may never become a top rider.

Positive. The successful rider has a positive, yet realistic attitude. I have noticed that riders who are frustrated often say, "I'm trying!" The word "trying" seems to carry with it an inference of expected defeat and usually results in difficult and frustrating work. I have found that if you start each session expecting the best but being mentally and physically prepared for the worst to happen, you will not be

surprised and will have most situations well in hand.

For example, if you are out riding alongside a road and you know your horse usually works slowly, calmly, and softly away from home, but usually gets somewhat heavier on the bit, stronger, and faster once you turn towards home, be prepared for this. Know ahead of time how to deal with such a tendency, but don't anticipate the behavior to the point that you are tense and inadvertently make the horse even stiffer on the way home. Just be prepared for what may come and use the techniques you have learned to deal with rushing if it occurs.

With such a horse you would probably canter away from home and only walk and perhaps trot toward home. Also you might stop the horse frequently for a few seconds both away from home and toward home. And just as you are about to turn

Riding along the road.

the horse toward home, you could stop him for a minute or so facing away from home, then turn him toward home and stand another minute or so, then begin walking. If your horse begins quickening, stop and sit calmly. Be sure your body is relaxed, centered and in control. Even if you have a strong mental picture of everything going fine, if you let your body exhibit tension or fear to the horse he will pick this up and respond in kind.

Have a positive self-image. Don't let your mental record play destructive phrases over and over again in your mind such as fat, skinny, clumsy, weak, uncoordinated, stupid. I have heard students proclaim these things about themselves yet my perception of them might have been sturdy, slender, relaxed, soft, inexperienced yet learning.

Patient. Becoming a good rider does not happen overnight. Because our world is filled with quick results and instant fixes, it is important to continually remind ourselves that the physical and mental development of an athlete takes time and effort.

A student that is very enthusiastic and excited about riding can sometimes get in a hurry without even realizing it. That is why it is ideal to work with a trusted mentor who can guide you toward your goal at an appropriate pace. The idea is not to find the quickest or easiest way to the top. Remember, it's not so important where you are going, but how you get there. Sometimes the longer, more thorough, but often more difficult way will make you a more solid and versatile rider.

If you have a problem with your riding at a basic level, have the patience and perseverance to work on it. If you don't, you can bet it will re-surface in future work and continually haunt you. In many cases it is best to resolve difficulties before moving on. This is called linear learning — taking one step at a time. However, if your problem brings your work to a standstill or even to regression, sometimes temporarily bypassing it and working on a slightly more advanced lesson will be beneficial. A good instructor will help you make such a distinction. Selecting and working with an instructor will be covered in Chapter 11.

Responsible. It is easy to accept the pleasure and luxury of riding. What is difficult for many people to accept is the responsibility for the associated tasks. The realities of horse ownership involve an investment of money, time, and hard work. Responsible horsekeeping includes soaping and oiling tack, shoveling manure, carrying water, moving hay bales, picking out feet, learning to punish your horse when he needs it as well as giving him a hug around his silky neck, riding up for your blue ribbon, and enjoying a gallop in the pasture with your friends. Sometimes the work portion of horsekeeping is lightly skipped over as it is not perceived to be as much "fun" as the riding. But if fun is defined as an overall sense of well-being, nothing equals the feeling of doing it all and doing it well and letting your horse's health, appearance, and performance show it. The dedicated rider approaches each part of the horse experience with equal enthusiasm.

Disciplined with good habits. If you own and care for your horse, no one is going to tell you when to clean his stall or pick out his hooves, and no one is going to give you a paycheck for doing it. That is why you must be self-disciplined and hard-working. Good habits and daily routines enable you to complete the important tasks on a regular basis and ensure your horse's health and well-being. Rather than waiting until the tack room looks like a cyclone hit it, tidy it daily. Rather than waiting until your horse's stall or pen disgusts you, pick it out regularly so that it smells fresh and is a healthy place for your horse to live. Similarly, keep your horses clipped, well shod, and neat at all times. You can train yourself to form good habits and have fun at the same time by barn charts and work boards, record books, and to-do lists. Every day, do one extra task that you can check off your list. Give your horse's mane and tail a thorough shampoo and moisturizing or take that western bridle apart and give the silver a thorough cleaning and polishing.

You can also develop a series of mental checklists to help you perform various tasks and maneuvers. For example, make up an acronym to remind yourself of the final tasks you must do before you leave

Develop good habits such as in-hand position.

the barn and head for the house. In order to leave the barn with a feeling of security, you must **WELD** it. You must turn off the **W**ater; **E**mpty the cart; turn off **L**ight; and close the **D**oor.

You can also train yourself to practice routine handling in a way that will benefit both you and your horse. Lead your horse through a gate or doorway in a specific fashion every time using key words to jog your memory as you work. For example: Halt, step forward, unlatch gate, open, walk forward, halt, turn on forehand, halt, walk forward, latch. Once you have formed a habit of doing something the right way, be consistent. It will instill confidence in yourself and in your horse. Developing and using mental checklists for barn tasks and ground handling will help you immensely when it comes time to develop some to use specifically in your riding.

Confident. It is no secret that confident body language goes a long way toward convincing a horse that you are in charge. Horses seem able instantly to pick up hesitation and confusion in the actions of their handlers and riders. Body language, after all, is the means horses use to establish and maintain their status in a

Confident attitude.

pecking order with other horses. Often a horse will try to bluff his way to a dominant position, but the decisive manner of a confident human or another horse will quickly put the bully in his place.

Steady, sure steps and deliberate movements give a horse the confidence that you know what you are doing, and this in turn puts him at ease. Horses are basically followers. When a horse finds that he can understand and rely on a human, he is cooperative and submissive.

Confident actions come from a mental certainty born from adequate knowledge, training, and experience. That is why it is a cardinal rule never to "overmount," that is, never to ride or handle a horse with which you do not feel confident. If you participate in riding at a level appropriate for your experience, you will operate in a confident manner.

Healthy ego. Perhaps it is the same in many other recreational endeavors, but it seems that among riders there are many overnight authorities. Bring up the subject of horses and often you'll find that anyone who has ever mounted and survived the ride considers himself experienced. Give the same person a few more hours in the saddle and he becomes a trainer. The next thing you know these "armchair horsemen" are spouting off advice to professionals.

Worse, from the horse's viewpoint, are those individuals who demonstrate to observers their "superiority" over horses by rude and inhumane tactics. These "show-offs" often create a confrontation with a horse and then try to win the battle through force. Unfortunately a novice rider may actually be impressed by such "horse breaking" and may not have enough experience to know that such behavior is not only unnecessary but undesirable. Anyone who uses a horse as

a prop to demonstrate his or her superiority makes a very poor teacher or role model.

On the other hand, a rider should not have a very fragile ego either. A very sensitive person can often take suggestions too personally. A rider should expect criticism because that is the way to learn. It is rare for an instructor to need to raise her voice if the rider approaches lessons with respect, humility, and perseverance. In order to make maximum progress in riding, the instructor must be able to work through the rider to mold the horse into the correct form. In an ideal situation, it's as if the instructor rides the horse through the student. This is only possible if the rider approaches the work with diligence, an open mind, and a healthy ego.

Strong mental control. The mind is a very powerful tool that can help or hinder physical performance in the saddle. How effective you are at controlling your mind depends on the relationship between your attention, anticipation, and arousal.

Attention. While you are riding, your mind must be on what you are doing. This can be extremely fatiguing, especially if you are not accustomed to focusing. But if you do not focus on your work, your horse will likely sense your lack of attention and will not perform obediently or at his peak level. Your attention is especially required at the initiation of each cue or set of aids such as a half-halt (check) or gait transition. As you are performing that movement, your attention will shift to the next part of your planned set of maneuvers. The rider who is able to maintain her attention better will be able to perform better. The greater a rider's skills, the less she will have to focus on the "now," the movement at hand. This will enable her to turn more of her attention to anticipation and planning of what is to come.

Anticipation. Anticipation is guessing

and planning what will occur and when and where it will occur. For example, if your horse has a tendency to let his shoulders fall out to the right when you are turning to the left, you can wait until he makes the mistake, then correct him, and then give him the aids to proceed correctly. Or you can plan an appropriate reminder for your horse ahead of time and use it just when he would be most likely to let his shoulder bulge to the right. This will minimize the number of stages in your line of communication and will greatly reduce the overall time it will take to perform the turn correctly. For anticipation to work effectively for you, however, it is important that you know your horse's most likely actions and what it takes to encourage or prevent them. If you pre-select the wrong response, you will have added several more links in the chain of commands and this will greatly increase the time it will take you to perform things correctly.

Your sense of timing will develop the more you ride. Practice will help you learn the optimal moment to apply a "preventative" aid. Experience will teach you that when you apply an aid depends on the position of the horse, your position in the saddle, the speed of the horse's movement, and the average response rate of the particular horse you are riding.

Arousal. Your state of arousal as you ride is the level of energy supplied to your body by your brain. It is a stimulation level that has a direct effect on the functioning of your attention and anticipation. Are you laid back or hyper-tense or somewhere in between? There is an ideal level that is appropriate for each situation, level of rider experience, and horse.

If you are operating at a very low level of arousal, any number of other stimuli

Attention, anticipation, arousal.

can steal your attention away from your riding. A paper blowing in the wind, a loud noise against the wall of your arena, or another horse running in a pasture can divert your (and your horse's) attention away from the work at hand. At a too-high level of arousal, your horse's tiniest movement (which might normally go unnoticed) can startle you into a state of "animal fright"; you can be easily distracted or derailed and you can lose control of the riding.

There is an optimal state of arousal that will allow you to focus on the important stimuli you are receiving from your horse's body. The level of arousal that is appropriate for you is governed by your personal anxiety level — the amount of stage fright that is necessary for you to give a peak performance without forgetting your lines! The more advanced you become as a rider, the more "psyched up" you may need to be in order to perform at your best.

Ability to concentrate. Your ability to concentrate will determine how well you can feel the feedback from your own body and your horse's as you ride. The more you concentrate, the more you will be able to remove yourself from the actual riding. You may find yourself observing your own overall performance as your body continues, almost as if on automatic pilot.

Concentration is especially necessary during a lesson so that you can hear and understand what your instructor is trying to tell you. Otherwise, valuable advice may be falling on deaf ears. It is not uncommon for riders of all levels to experience lesson anxiety that makes the instructor sound as if she is talking in a foreign language or using words that have no meaning.

You should not ride if you are preoccupied with other thoughts. Besides being non-productive, this can also be dangerous. You should learn to turn off your internal dialogue — that voice inside that keeps reminding you about your job, your family, your schoolwork, the disagreement you had with your spouse or best friend, the disturbing phone call you received last night, an upcoming important business deal. Becoming consumed by matters of the "self" can be very limiting. You need to release your concerns so that you can become absorbed in your riding. Turning off the internal dialogue will allow you to experience a new level of awareness. Rather than being caught up in the details of your "self," let go, and ride unencumbered.

Focusing will help you concentrate. To ride safely and effectively, you must stay alert and focused on yourself and the horse. The phrase "Be here now" may help remind you to forget the past and future of your self and allow you to be immersed in the now and in the act of riding. If you let yourself become absorbed in your work as you ride, you may receive a bonus after

Turn off your inner voice.

your riding session. You might find that your concerns are not so mountainous after all, and the solution to a problem may suddenly appear in your mind.

Ability to relax and control anxiety. Riding requires a high degree of mental alertness but performance deteriorates when anxiety reaches the point of fragmentation. The ideal state is one of keen, "peaked," but relaxed attention. Some riders perform well in a high state of arousal, while others do their best in a quiet, relaxed state of mind. And although a particular rider may more easily learn something new in a relaxed state, she may need to increase her state of arousal to a higher degree to perform a familiar task. This means it is beneficial to learn how to increase or decrease your anxiety level automatically. When you are able to do that, you are basically changing the level of adrenaline in your bloodstream.

Adrenaline is a hormone that is secreted during stress. As soon as it is released into the blood, your metabolism increases characterized by a rise in your blood pressure, oxygen consumption, and an increase in the activity of your central nervous system. Adrenaline reduces the blood flow to your gut and belly but increases the blood flow to your skeletal muscles. This increases the stretch of your muscular contractions and results in an increase in strength, alertness, coordination, and reflex action. Therefore, you must have a controlled use for all of this new power or you may become a bundle of energy with no place to go!

You also must be able to learn how to turn down your anxiety meter. That's where relaxation comes in. One way to relax is consciously to build tension to its highest state and then release it. For example, if you find that you are so tense that your thighs squeeze together and lift you right out of the saddle, then inhale and turn those thighs into hard rocks with the strongest muscle contraction you can muster. Then totally let the contraction go as you exhale. This will give you a physiological basis for relaxation. There will be more on this in Chapter 5, "Exercise." Remember, however, that the essential key to physical relaxation is having a strong mental focus.

Emotionally stable. All other factors being equal, your emotions can tip the scales toward your success or failure in riding. If you face your emotions, you can use them as an asset to your development rather than a hindrance. Your emotions are largely governed by chemical changes that occur in your body. Some biological effects, such as those caused by adolescence or by premenstrual syndrome (PMS), are rather fixed, even if only temporarily, and you must learn to deal with them. Other undesirable chemical changes are caused by habits that can be changed such as poor diet and drug and alcohol use.

To use premenstrual syndrome as an example of how emotional mood swings can affect riding, it has been found that some women experience a chemical change and a resulting mood shift one to two weeks before menstruation. About three-quarters of the women affected by PMS experience nervous tension, irritability, and anxiety. Other symptoms include water retention, swelling, extreme cravings for sweet or salty foods, depression (including crying), confusion, and forgetfulness. All of these conditions can undermine riding. PMS symptoms have been found to be aggravated by poor nutrition and alleviated by proper nutrition. It has been suggested that Vitamins B, C, and E and the minerals calcium, magnesium, and zinc, as well as other nutrients, are helpful in diminishing the undesirable effects of PMS.

Learning what affects your emotions and how to control them is essential to your development as a rider. Sports researchers have noted the following facts related to the athlete's emotions. The fear of failing can contribute to motivation. Moderate emotional excitement tends to be an aid in extending physical exertion beyond normal capacities; however, extreme emotional stimulus may produce tension or "buck fever," and can have a negative effect on both learning and performance. Too many emotional peaks in one period of time may contribute to staleness.

Honest. Through the entire process of learning a new skill, it is important to be honest and have the highest ethics. Some folks try to do things the quickest and easiest way, and that's fine when it's a bale of hay that needs to be moved. But when the issue is self-development it would make no sense to bypass the basics or try to buy your way to the top. There is no substitute for the hard work and time that is required to become a rider.

Especially as a rider becomes more experienced and acquires lofty goals, the temptation to take shortcuts is there. There are many clever little ways to "cheat" or fool yourself, an instructor, or a judge into thinking that you are really doing a first-rate job of riding a horse. Short-cuts include constant manipulation of the reins in an attempt to make a horse look under control, or using a suede-seated saddle to help the rider "stick" to the horse. These are Band-Aids. They might cover up a problem for a short while but in the long term, there is still a hole in your riding.

A person dangerously sabotages her success and personal ethics if she thinks that anything is fair that she can get away with. You should never depend on an official or an instructor to "police" your actions. Many sports today are based on the strategic use of fouls, where coaches plan to use as many as they can get away with. That strategy is inappropriate for riding. After all, the initial reason most people pursue riding is for the feeling, for the experience of working with the horse.

Therefore it is the quality of the moment that is most important; the score at the end should be of secondary importance.

Balanced. Balance refers to much more than sitting correctly on your seat bones and keeping your upper body from toppling over. It goes far beyond the correct and effective application of aids. Total balance in riding includes a strong internal focus which is largely dependent on a person's awareness, concentration, and confidence. While it is true that a rider may possess the techniques to ride an advanced horse, the physical means are not the total picture. There is a certain internal binding force that ties everything together into a harmonious package. A rider that is mentally, emotionally, and physically in balance has the greatest chance for success.

Some students of riding bury themselves in intricate theory, thinking that book knowledge alone will make them good riders. Although a certain amount of the right kind of information is essential to understand the mechanics of riding, too great an emphasis on information and theories can actually encumber a rider. Learning to ride depends more on discovering how it feels to ride.

CHAPTER 2

VISUALIZATION

Athletes in many sports use visualization to dramatically improve performance by developing a new internal picture. Some of the ways you can build your visual "library" are by looking at photos and videotapes, and by watching your instructor ride and give other people lessons. Visualization is a strong mental tool that will allow you to become aware of your body, deal with trouble spots, fine-tune your position, and develop confidence and poise.

Effective visualization starts with astute observation and an altered way of seeing. The keen rider learns not only to look, but also to see. She sees in two ways: with a concentrated focus and with a "wide-angle lens." When she wants to see actual details she uses the concentrated focus, such as when she intently watches another person's aids as if accumulating the pieces to a puzzle. Or she may become a hoofprint detective, reading a horse's fresh tracks on a roadside to determine in which gait the horse was traveling, how straight he was traveling, if the hind feet

were overstepping the fronts, and so on.

When this keen rider is mounted she uses her "wide-angle lens." This allows her to take in the big picture without concentrating fixedly on one portion of her riding. She senses the relationship of the many body parts and movements — both human and equine — that are involved in a particular maneuver.

The Ki

In order to coordinate all of your aids in a smooth and harmonious fashion, you must have a strong central focus. This focusing or centering is referred to as the ki by aikido masters. The ki is the life energy, which has the potential to give body movements a controlled power. It is a relaxed energy force that allows you to perform effectively yet remain quiet and sensitive. Open your mind to the development of this energy body. Does such terminology have the flavor of an Eastern philosophy? Most definitely. How does that make you feel? Are you immediately saying, "But I need to learn the correct aids for a right lead lope depart or a shoulder-in!" Put that, other rational thoughts, and your internal dialogue aside and let the ki unlock the door to your riding awareness.

Concentration cannot be achieved simply by willing it; you must learn how to tune in to yourself and tune out extraneous activity. This will allow you to be totally engaged and involved in your work. You will have developed a state of riding consciousness. Rather than thinking about the various parts of the performance, you experience the activity in total. Visualization helps you to that end.

VISUALIZATION

Visualization is a combination of technique and attitude that allows you to develop mental images of your improved performance. Sports psychologists have found that when athletes run mental motion pictures though their minds, they are, in effect, training their bodies to improve skills, enhance coordination and timing, and increase confidence. And one of the greatest features of this technique is that you can practice your riding in your mind almost any time, anywhere, and be improving.

Now Showing at the Right Brain Theater

Visualization takes place in the right hemisphere of your brain, the portion of your brain that is responsible for the coordination of visual perception and the kinesthetic sense. The right brain stores visual images as well as those of color, texture, and sound. It is the center of your receptive, intuitive self and your musical abilities. It is also where your realizations take place.

Developing the right side of your brain will allow you to be more in tune with what you see and feel and how it relates to the spatial arrangement of your body parts and your horse's. The more you allow your right brain to take over when you ride, the more highly developed this sense will become and the more natural your balance and movements will be. Riding in large, open spaces, outside of the ring, often helps you find and develop your right brain, because you aren't constrained by arena walls or rails and aren't forced to make frequent turns at specific places.

Your left brain, meanwhile, is busy storing theory and logically solving riding

Ride in large open spaces to develop your overall sense of position.

problems such as developing exercises to square up your shoulders or understanding the sequence of aids for the turnaround. The left hemisphere of your brain is the area where sequential information is stored. How-to and step-by-step procedures as well as logical and rational thought processes go on here. Information is stored in segments in your left brain. When you use your left brain you figure things out part by part, in contrast to the overview taken by your right brain.

Theory is necessary. But it is best to do your intellectualizing and procedural development in between your saddle times. Once you mount, leave the internal intellectual dialogue behind and concentrate on listening to your body and watching the pictures in your mind. You will find that your body will elicit a correct response from your horse more automatically if you are not constrained by the intellectual aspects of riding during your actual ride. Many top riders ride horses instinctively and cannot explain the theory. They have grown up seeing horses ridden correctly and it is part of their sensory awareness. So, read before you ride, watch good riders, and then let your right brain be your guide when you are in the saddle.

Take a Good Look at Yourself

Like any system of improvement, visualization first requires a detailed self-evaluation. Where are you starting in terms of your present riding ability? Once you have established that, you need to decide what your goals are and how you will reach them. Rather than set vague, long-term goals, it is best to set smaller, more specific, achievable goals that can

be reached within about six months. Then you can set more new goals and more new goals and more new goals, and visualization will help you reach them all.

Your self-evaluation is best done with the aid of photos and videotapes. A set of photographs taken every six months will provide you with a convenient record of your progress. So that you get the most benefit from the photos, dress for the occasion. Wear well-fitting clothes, not designer trousers, loose blousey shirts, or long baggy T-shirts. Clean lines around your shoulders, waist, and legs will highlight the tendencies in your position. Tops with horizontal stripes are helpful, provided you have them on perfectly straight. Stripes can be a dramatic indicator of your shoulder and upper body position. If a striped shirt is twisted, however, it will present a deceptive picture of your left-to-right balance.

Horizontal stripes can help in assessing position.

If you wear a western hat or helmet, placed correctly on your head, its brim will indicate whether you tend to tilt your head or to look down as you ride. Both habits can throw you and your horse off balance. Wear gloves that will highlight your hands. If you are riding a dark horse, wear grey or tan gloves; with a light horse, wear dark gloves. If your lower leg position is a trouble spot for you and you feel your dark pants and boots will be lost on a dark saddle or horse, wrap the center of your calf, your ankle, or your stirrup in white or bright colored tape so you can more easily see what your leg is doing.

A photograph can only catch your action during one split second and may therefore be unrepresentatively flattering or unflattering. That is why I suggest having someone shoot at least a 24-exposure roll of film during one of your typical rides. Ask your photographer to use the following plan as a guideline. The resulting pictures will give you a good idea of your overall riding skills.

Photo plan. For the most benefit, you need to see yourself from all sides at all gaits. The most useful views are taken when the horse is working in a straight line: the side shot, the front shot, and the rear shot. You and your horse should fill up as much of the frame as possible to give you the greatest amount of detail. This usually requires that the camera be equipped with a telephoto zoom lens so the photographer can get a close-up of you from any part of the arena.

Techniques will vary depending on the height of the horse and the photographer, but in general, the camera should be pointed somewhere below the rider's waist. For the side shot, the camera should be aimed directly at the rider's thigh. This will result in the profile photo or "rail shot" that is essential for evaluating your overall position. The front shot should show both of your legs, so the camera is usually pointed at the midpoint of the horse's neck or chest. For similar reasons, the rear shot should be aimed directly at the center of your horse's tail. Have a side shot, a front shot, and a rear shot taken at each gait in each direction. This will use up eighteen of your photos, leaving six for miscellaneous views. Photos of circle or turn work will reveal whether you have a hollow side or a collapsed hip.

Using your photographs for self-evaluation. When you get the photos back, if you ride western or dressage, look at the side shots using a ruler or straight edge to help you determine if there is a straight line from your ear through your shoulder through your hip to the back of your heel. Where is your hand in all of the photos? Is it consistently in a 4 x 4-inch box in the vicinity of the horse's withers? Look at the front and rear shots to see if your legs hang unevenly, if one shoulder is higher than another, or if one hip is farther forward than the other, causing you to sit your horse diagonally.

If you're like most people, the photos will look the worst to you the first time you see them. Don't throw them away! After a few days or a week, look at them again. Choose one or two of the photos in which you are riding your best and one or two of the photos that really show your weak points.

Now find photographs of a top rider whose physique approximates yours. We'll call her TR. Use a photograph of TR with her horse standing still and some showing active riding. Place the photos alongside your best and worst photos. Create a list of the attributes that make TR a top rider. I am not talking about the horse, tack, or clothing, but the effectiveness of her body parts and overall skills. Choose two or three specific attainable goals from your list and put a big star by them. For example, if you notice that TR uses a longer stirrup, and you feel it adds to her effectiveness, then that may be a goal for you to work toward. But it's not as simple as just using a different hole on your stirrup leathers. Examine the photos to see how TR can ride so well with longer stirrups, and you will probably discover that she has a much deeper seat than you do. The deep seat will then become the means for you to reach your goal of riding with longer stirrups.

Now look again at the photos of yourself and your role model TR. Close your eyes and picture your face on TR's body. You may find yourself all of a sudden sitting up much straighter in your chair with a smile spreading across your face. Let TR's body gradually turn into yours. As you keep TR's good position, go through the various parts of your body in the saddle and note how they feel. Maybe your new, improved mental picture seems very tall in the saddle. Experiment with ways to stretch your spine in order to lengthen your upper body. Maybe TR's seat was indistinguishable from the saddle — as if it had melted into it. Experience a warm sensation in the muscles of your seat, thighs, and back that allow you to really sit down on your horse. Focus on the body parts that you most want to improve and add to your new, improved mental image. If you lose the image, open your eyes, look at TR, and begin again. After you have finished, file away your photos so that you can use them to compare to the next batch of photos you will take.

LIGHTS, CAMERA, ACTION

Now you're ready for mental action. The next stage of building your mental imagery is a dynamic version of the photo session, this time using videotapes. Videotapes of your lessons, show ring performances, or daily training sessions will allow you to see the tendencies in your riding that need the most attention. You will get a sense of your consistency, balance, timing, coordination, and ability to follow the movement of the horse.

A properly made videotape will be very useful, but a haphazardly made film can be counter-productive. The video camera should be mounted on a tripod so that the panning action is smooth as the camera operator follows you at the various gaits. It is rarely possible to get a steady, useful film when the camera operator holds the camera on his shoulder. More often than not such a film does a better job of registering the cameraman's breathing, throat clearing, and position shifting than your riding! Follow the attire guidelines for the photo session to make your videos most useful. It is best not to wear red or orange when you are going to be videotaped, as those colors tend to "bleed," making you look blurred with fuzzy edges. Very bright light and white clothing sometimes blind the camera and may result in less contrast or detail. If you are taking a lesson and want your instructor's comments to be keyed directly to your movements, ask your instructor to wear a remote microphone that is linked to the camera.

Developing a Deep Seat

Focusing (or centering as it is sometimes called) allows you to lower your center of gravity and feel heavier so that you can sit deeper in the saddle — a goal for any rider. You can get in touch with your energy center through exercises that help you become aware of your center of gravity. Once you focus on your physical center, you will find it easier to develop a sense of your energy center. This will help you to become a balanced rider.

Your center of gravity is located about one or two inches below your navel and deep within your abdomen. You might become more acutely aware of your center by pressing firmly on the area below your navel and then gradually releasing the pressure. Often the sensation remains. Closing your eyes sometimes helps you sense your center better than if your eyes were open to "distractions." But your center is not merely this anatomical spot. The center of gravity, termed hara by the Japanese, is the core or essence of your being, through which you can experience an uninterrupted flow of coordinated energy, both physical and mental. By breathing properly, you should be able to make your physical center of gravity seem lower and heavier. If you are on a horse when you do this, it results in a deeper seat, one of the essentials of good riding.

Evaluating the videotape. As you view the videotape, take notes on your strong points and the areas that need improvement. As your observations accumulate, your mind will subconsciously sort and store the pictures it sees into two main stacks: CONTINUE DOING THIS and CHANGE THIS.

Once you have had a chance to view yourself in motion, watch a tape of TR (or another high-caliber rider) at work. When you get to a portion of the tape where TR is performing a maneuver that causes you difficulty, play the tape on slow motion, or freeze the frame and advance it one frame at a time so that you can sort out each step of the overall maneuver. Note when a shift in balance occurs and how that affects the horse. Keep an eye on the position of TR's hands. Watch the action of TR's pelvis and lower back and the action of the horse as a result of her deep seat.

Think about TR's performance for a day or so, then go back to one or two key portions of the tape that made the most impression on you and watch them several times. Turn off the tape and close your

Use videotapes to help you visualize your way to better riding.

eyes in a quiet room with no distractions. Replay in your mind's eye what you just saw on tape. Then while you let TR continue riding in your mind, let her face fade out and replace it with your own features and just keep on riding. Let her clothing gradually turn into your shirt and jeans and keep riding. Watch as her horse's star changes into your horse's blaze. Then, with your eyes still closed, ride right out of the videotape setting and into your own arena. Repeat the key maneuvers several times. If you make mistakes, go on and correct them in your mind as TR would. Work as long as you feel comfortable and can stay focused. Then gradually bring your horse to a walk and open your eyes so you can lead him back to the barn!

When you took over the reins from TR, you likely had your very best ride to date. And whether you know it or not, your nerves and muscles were learning how to direct your body in their new motor skill patterns while you sat in your chair riding in your mind. The benefit of your mental ride will show the next time you mount up.

After you've had a chance to actually ride your horse, go back to TR's tape. Maybe you'll only have to watch it for a few minutes before you feel the urge to close your eyes and get into the mental saddle once again. And soon, you will not even have to turn on the video. You will be able to close your eyes anywhere, anytime, and bring that picture back to mind. Some of the most logical times to call your own top ride movie to mind are just before a horse show class, a lesson, or before your daily training sessions. Each day as you warm up, close your eyes for a moment, turn on your mental tape, and picture your way to better riding. Soon you will find that you ride in an altered state.

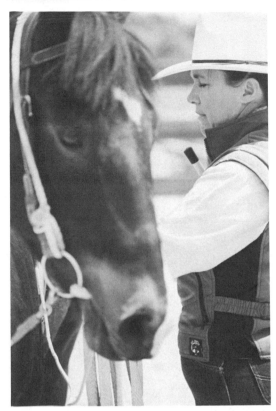

Close your eyes and form a picture in your mind of how things will go.

The altered state. Sometimes riding can feel almost like a hypnotic dream state. You may be performing very intricate and demanding work with your horse, yet you feel peaceful and euphoric. It is almost as if you perform inside a capsule or cocoon that moves around with you. This aura or energy field surrounds both you and your horse when you are working in harmony. When you ride in an altered state, there is no need to time your ride to be sure that your horse is receiving his minimum amount of exercise time. Time flies! One hour of focused work can seem like only fifteen minutes to you. Yet when you are not focused, a fifteen-minute ride can seem to last FOREVER.

The tone during an altered state is very harmonious because you use the optimal amount of energy for each of the

The altered state.

future by slamming the door on it suddenly or forcing yourself to snap out of it. Instead make the transition into ordinary reality by slowly untacking and caring for your horse. Savor the feeling of quiet intensity and hold the picture in your mind for a preview of your next ride.

FROG TOES AND LEAKING PIPES

Another way you can utilize visualization or mental imagery is to look at a picture of something very separate from horses to help you overcome a particular problem in riding. For example, if one of the reasons your right leg is always drawn up, shorter than your left (causing you to sit with a

aids and you use the aids in smooth coordination. Your horse shows no signs of nervous fatigue and although he may be working very hard on difficult maneuvers, the motion is smooth and gives you the sensation that you are floating through the work. Both you and your horse breathe in low, long, rhythmic breaths. In this dreamlike euphoric state, the overall performance is very consistent, allowing you to notice minute details about your position. Changes are subtle and effective.

When you finish the active portion of such a ride, you may feel a deep sense of satisfaction and elation as you begin to cool out your horse. You also may find that you cannot talk in the clear, rational manner that you usually do — but don't let this alarm you. You have experienced a part of your being that you probably do not let come out during your normal daily activities. Don't discourage this state of consciousness from reappearing in the

Use mental imagery to help you with problems.

collapsed hip), is because your toes are clenched in a tight ball inside your boot, you need to find a way to relax your foot. You want to spread your toes broadly inside your boot so that they cover your stirrup tread and provide a secure, relaxed base for your leg. Perhaps a photo of a tree frog will help you remember to unclench your foot. Instead of going through intellectual, physiological explanations of why and how you should relax your foot, just let the picture of the tree frog flash through your mind. The correction will take place quickly and simply and be lots more fun. If you use mind pictures, I'll bet you won't be able to keep from chuckling to yourself as you ride.

Visualization can help you understand your horse's movement, too. Choose a vivid picture image. For example, if you have a technical or mechanical mind, you might think of the energy flow around a horse's body and his overall position as a system of flexible plumbing pipes and joints. As your horse moves, you can sense the flow of water through the pipes. As long as the horse is moving forward freely and straight, the water flows evenly through the pipeline system.

When a problem occurs, such as the horse traveling crooked or throwing its head up abruptly, one of the fittings may loosen, spring a leak, or even break, and water may begin to gush out of the horse's shoulder or his neck. Besides "getting wet," you will notice that your ride becomes less comfortable, and it will become more difficult to sit properly when

Imagine that you have sponges in your hands as you apply your rein aids.

your horse is "leaking." As you make the proper adjustments in your body and the horse's body, the leaking slows down and finally stops. Your mental picture goal, in this case, is to ride so that the water courses freely throughout the horse. Soaked from the last flood, you'll become acutely aware of pin-hole leaks and take care of them before they become a major problem. You'll find yourself sensing areas of your horse's body where water tends to back up, and you'll automatically do the right thing with your body to see that the water pressure doesn't build to a dangerous level. Such mental pictures help you sense what is happening overall without you concentrating so much on riding techniques that they keep you from riding your best.

CHAPTER 3

GOALS

Your goals are your targets, the end results toward which you aim your actions. Goals increase incentive and fuel personal motivation. Horse competitions have inherent goals: the fastest time, the highest jump, the blue ribbon ride. Behind each successful performance, however, are many sub-goals and objectives. Even if you are not competitively oriented, you will set goals in a fashion similar to the competitive rider. Most riders, whether they know it or not, have immediate goals, short-term goals, and lifetime goals.

Immediate goals are usually very specific and measurable, such as "I will ride my horse between cones H and E" or "I will canter my horse on the left lead at marker A."

Short-term goals can be thought of as those that are attainable within a six-month period. Depending on a rider's present level of experience, her short-term goals might include "I will ride a Fourth-Level dressage test at a recognized show" or "I will go on a two-hour trail ride" or

"I will tack up and ride a horse without assistance."

Lifetime goals are understandably broader than immediate or short-term goals. They are often vague, even though they should not be: "I will become a good rider" or "I will participate in a horse show" are common lifetime goals. Adding specifics makes a goal more measurable: "I will go on the Sierra two-week horseback camping trip before I am forty" or "I will make the United States Equestrian Team by 1996."

As you prepare to set your goals, be sure they are based on YOU becoming better than YOU presently are rather than you becoming better than someone else. For you to win, it is not necessary that someone else loses. And just because someone else is successful does not mean that you have failed. Other people have different bodies, horses, schedules, financial means, facilities, abilities, and instructors than you do. If you compare yourself to others, you can become frustrated, discouraged, unrealistic, and as a result unfair to both yourself and your horse. Or you can develop a dangerous feeling of superiority, which can block your further development.

Learn to be satisfied with YOUR progress. Although the greatest amount of learning takes place at the beginning of a new endeavor, some horses (because of their conformation or training) will allow you to make only very minor, slow progress with your riding. And the more advanced you become as a rider, the slower your progress may tend to be. So the best thing to do is compare yourself today to yourself yesterday, keeping a standard in mind, and note the progress you make.

If other people (such as parents or spouse) will be involved somehow in the process of you reaching your goal, be sure

This jump is a concrete goal.

to get agreement from them. This is crucial if they will be required to invest money or time in your quest or will be affected by your money and time expenditures.

Write down your goals in finely focused and specifically described terms. Broad, vague goals are less likely to be reached, as it is difficult to determine when you have reached them. It is important to write down your goals rather than just think about them. People with explicit, written goals usually outperform those without them.

Often a person is hesitant to write down goals through fear of failure or fear of success. If she writes down a goal and *doesn't* reach it, she feels she has failed. But what failure often means is that a goal was not properly set. If she writes down a goal and does meet it, on the other hand, other people tend to modify their expectations of her. She may feel this puts pressure on her to maintain her higher level of achievement. Or it may make her feel she has to explain why she may wish to change her goal mid-course. Such feelings are a result of setting a goal to fulfil someone else's wishes rather than her own.

Markers can help you define your goals specifically.

SETTING GOALS

It is crucial that you take time and care when setting your goals. Your success depends on it. To begin setting a goal, write down your overall goal:

To compete respectably in a trail class at a horse show.

To keep your goal as detailed as possible, you will need to define "respectably":

A respectable performance would be to perform the class requirements with prescribed tack and equipment, without any refusals or knocking over any equipment.

To close in on your goal, you should list its sub-goals:

The class requirements are:
- **Working a gate**
- **Crossing a bridge**
- **Crossing logs**
- **Putting on a slicker**
- **Opening a mailbox and taking out newspaper**
- **Dismounting, ground tie, remounting**
- **Sidepassing over a pole**

To narrow the focus of your goal even further, you should choose one sub-goal and list its components:

To work a gate:
- **Ride horse forward**
- **Halt**
- **Sidepass left and right, holding the gate**
- **Turn on the forehand, holding the gate**
- **Back, holding the gate**
- **Work reins and gate with either hand**

Working a gate is a sub-goal of trail riding and contains many objectives.

Now you must choose one component of the sub-goal and define your objectives for it. Your objectives are your means for reaching the sub-goal. Write them in the present tense in a positive tone. Your objectives will usually fall into four categories:

- Attitudes you need to develop
- Habits you need to develop
- Skills you need to develop
- Outside influences you need to control.

Your attitudes include your desire to reach your goal, how you will feel when you reach your goal, and what it looks like in your mind's eye as you reach your goal.

Your habits include a mental rehearsal of how you will prepare for your goal, what you will do to reach your goal, what you will not do in order to reach your goal, and what you will do if things go wrong.

Your skill development includes body awareness, exercises, and motor skills that you will practice both on and off the horse.

Outside influences include proper equipment, footing, and factors in the external environment.

To continue with the example of working the gate, here is how you might define your objectives for the sub-goal of sidepassing left and right while holding the gate.

- **Attitude.** I feel a cooperative smoothness as my horse and I move sideways and the gate swings quietly on its hinges. I see my horse putting me in position to operate the gate easily without letting go of it.

- **Habit.** Every time I see an opportunity to work a safe gate from horseback, I do it. I accustom my horse to a new gate by walking up to it and stopping for a moment. I'm never in a hurry when I work a gate. As I work a gate, I pause regularly so that my horse does not become anticipatory. If my horse pushes through the gate or swerves suddenly so that I lose my hold on the gate, I stop, regain my composure and control, and begin again.

- **Skills.** I do 20 leg lifts every day so that I am able to use independent leg cues required for the sidepass. I practice holding the reins in both my left hand and my right hand.

- **Outside influences.** I check to see that the gate swings freely, that I can reach the latch from horseback, that the latch is easily operable with one hand, and that there are no rocks or holes near the gate.

Be **SMART** when you set your goals and objectives. Keep them

- **S**pecific
- **M**easurable
- **A**ttainable
- **R**elevant
- **T**imeable

For example, if you want to participate in a hunter hack class, you need to be able to canter your horse over a 2-foot jump. Is that a **SMART** goal?

- **S**pecific: Getting over a 2-foot jump.
- **M**easurable: You either get over the jump or you don't.
- **A**ttainable: Almost any horse can jump 2 feet.
- **R**elevant: It is essential for the class and is good training for any horse and rider.
- **T**imeable: Within two months I will be able to accomplish this goal. Yes, this is a **SMART** goal.

PLAN THE WORK, THEN WORK THE PLAN

Once you have zeroed in on your goals, start heading toward them. It is imperative that you believe in what you are doing. Begin now, not later this afternoon, tomorrow, or after New Year's. Procrastinate and you vegetate. If you need to bolster your incentive, itemize the benefits that reaching your goal will bring.

Some people find positive statements useful for keeping on track. A positive statement about your objectives can be thought, spoken, or written on a regular basis as you work toward your goal. Positive statements are part of the private conversations you have with yourself. But beware: if you choose to make such statements out loud to your friends, you may be considered a braggart! You need to devise, plant, and sustain positive statements to fit your personal beliefs and feelings. Some statements tend to be businesslike:

I drive with my seat and lower legs and my horse goes over the jump.

while others are introspective:

As I ride, I feel the productive energy flowing in and the negative energy flowing out.

If you find that your mind has negative responses when you try to use positive statements, try to find out what stands between you and your goal. Often a negative reaction will vividly point out the obstacles standing in your way. For example, if using a positive statement makes you say, "This is for sissies!" it may indicate you need a gentler approach to training and riding horses. On the other hand, if a statement makes you feel like a drill sergeant, perhaps you need more organization and discipline in your riding. Your tendencies will be revealed in your reactions to positive statements.

As you work toward your goals, allow yourself to be influenced only by healthy, positive, supportive people. Misery loves company, and people often unintentionally try to drag other people down to their level of frustration or dissatisfaction. Stay focused and positive and plan to work mostly on your own. If you are lucky enough to have a friend who is also positive and directed, be supportive of him or her.

Anticipate problems that may occur along the way to your goal. Carry the solutions around in your back pocket so that you can institute them quickly. You will need to know how to make sound decisions when problems surface. When you make a decision, you determine what type of action is to be taken and when. In order to make an effective decision you must be informed, observant, expedient but not hasty, and ready to take action. The action may be designed to correct a problem, prevent a problem, or temporarily bring a problem to a closure until the proper action can be determined. Or an action can replace Plan A, a previous decision and action, with Plan B, a more-likely-to-succeed solution. Rather than be carried along for the ride, make decisive steps as you work toward your objectives.

It helps if you keep records so that you can evaluate where you have been, where you are, and where you are headed. Photographs, videotapes, and journals can all provide specific benchmarks for you to use periodically in your self-evaluations. Be observant and flexible so that as you evaluate your progress you can make necessary changes and restructure your program. If in your review you find that your original plan is no longer appropriate, revise it and move on to a new plan. Be sure that when it is necessary, you take time out for mental and physical recharging so that you come back with renewed energy and enthusiasm.

When asked by their instructors, students state that one of the most common limiting factors in achieving their goals is the lack of time. It is a fact that riding and caring for horses requires a good deal of time. Often, though, when a person feels she lacks time, what she really lacks is quality time. Quality time is characterized by focused, productive work. It is not how much you get done that is important but how well you do what you do.

To ensure that the time with your horses is quality time, don't be in a hurry as you work with them. Your goal-setting exercises will help you determine what to do. Do first things first. Focus on what you are doing so that you do things right the first time and don't create large, time-consuming problems. Finish what you start. Be orderly; it will save you immeasurable time. Have a place for everything and keep everything in its place. Use written and mental checklists to give you a measurable sense of completion.

To get your horse to cross water, keep a positive attitude but anticipate problems that might occur and have solutions ready.

PROBLEMS RIDERS COMMONLY FACE

Stress. Stress is a demand for adaptation. A certain amount of mental and physical stress is necessary for the development of the rider. After healthy stress, such as a soreness in the thighs, a rest period will allow a rider to rejuvenate. Yet when the amount or frequency of stress exceeds a healthy level, a rider's reasoning, problem-solving ability, and effectiveness decline. An overdose of stress may cause irreversible damage to the physical and emotional reserves of a rider. Get in touch with your stress barometer so that you know how much is just enough and how much is too much.

Tension and fear. Horses are very sensitive and can easily detect a fearful or tense rider. Every horse's reaction to tension and fear is slightly different but may include biting, kicking, running, shying, rearing, stiffening, and balking. Experiences causing fear and tension in a rider can range from a violent accident caused by broken tack to a momentary loss of balance resulting in the rider quietly sliding off the horse. Read Chapter 10 on Safety to be sure you are following recommended guidelines for safe horse handling and riding. And ask yourself these questions related to your fears: Are you physically or mentally unable to perform what you are attempting? Do you have an instructor who can give you feedback on your riding? Are you practicing regularly? Have you been badly hurt? Have you felt embarrassed in front of others? Are you keeping your progress in riding in proper perspective to the rest of your life?

Plateaus. What kind of picture comes to your mind when someone says their

This novice was afraid of being bitten and was confused and embarrassed about the procedure for bridling. She arrived for her riding lesson on this day with successful bridling as her primary goal. She listened carefully to step-by-step instructions and, with the assistance of a well-mannered horse, achieved her goal. These photos chronicle her very first bridling experience.

riding is on a plateau? A plateau can be a resting spot where you rejuvenate and regroup, or a level, clear, elevated place to hone your skills by practicing what you already know. Or it can be a rut in which you are frustratingly stuck.

There is nothing wrong with heading on up to a plateau for the specific purpose of taking an extended break. In fact, knowledgeable horsemen can sense when it is time to schedule in an official rest period for their own and their horse's benefit. Similarly, if you see that your schedule will cause your upcoming riding sessions to be irregular and of poor quality, plan a period of R and R. Hang up the bridle and turn your horse out for a week or so. When it's time to resume, you will most likely find that both you and your horse approach the work with more gusto and interest. To preserve interest and prevent staleness, vary the work routine to include arena schooling, trail riding, and road riding.

Many beginning and intermediate riders stay on a plateau, practicing a certain level of riding for a seemingly endless period of time. They practice the things that come easily. They may adopt a casual overall riding style or they may perform over and over again certain maneuvers that their horse knows well. It is common to see a horse and rider loping round and round an arena as if this is all they know how to do. It may be that the horse's transitions are not really smooth yet, so the rider prefers to stay in one gait for a prolonged period of time. On the other hand, a novice rider may practice a particular maneuver over and over again, more times than are necessary or good for the horse, simply because the horse knows how to do that maneuver well. This is a natural tendency for developing riders and it does serve an important purpose.

Doing those things that come easily and can be performed well allows a rider to develop skills and a feeling for the aids. It is especially important for a rider to spend many hours at the trot and canter when developing a seat for any style of riding.

If a rider works repetitiously on a plateau, however, she may begin forming ruts. Anyone who has ridden in the tracks along the rail of an arena knows what I mean — once you have slipped into the ruts, it is difficult to get out of them, somewhat like slot car tracks. These ruts can be symbolic of the mental stagnation you and your horse may be feeling. Repeating the same maneuver over and over again, especially if the horse is doing it well, can bore the horse, dull his responsiveness and willingness, and keep you from progressing.

When a rider repeats the same things over and over again because she does not know how to solve a problem or does not know what to work on next, then a plateau can represent a frustrating standstill. But remember, frustration can motivate you to learn and improve.

As you progress in your riding, you will find yourself spending less and less time on the familiar, comfortable things and more time on things that challenge both you and your horse. You may focus on advanced maneuvers or a more specialized or formal style of riding.

Sometimes a rider will find herself on a plateau of staleness. A particularly intense rider who spends a great deal of time involved with her sport may experience a general fatigue due to overtraining. This may cause her to lose interest or enthusiasm, have slow reflexes, and be tired or irritable. Although poor diet, inadequate sleep, home worries, or excessive school or work load may also cause staleness, the major cause is usually excess emotional tension due to intensity of involvement with riding. The best way to come out of a stale period is to participate in a diversionary playful activity — perhaps an activity that you have never pursued before and one that has nothing to do with horses. When you return to your riding, bring with you that sense of play. If staleness presents itself regularly, it may be a sign that you need to lighten the intensity of the riding goals you have set.

Sometimes following an accident or a particularly frustrating riding session, a rider will simply lose all interest in anything to do with horses and will withdraw. Or the rider may choose never to ride a particular horse again. Most riders in such a situation just need a period of rest and rejuvenation. In some instances, however, the rider mentally builds a wall between herself and riding. The longer the period of non-riding, the more difficult it will be to get over that wall. During such a time, the rider should reevaluate her goals and then re-enter riding with safety and enjoyment of foremost importance.

SECTION TWO

PHYSICAL DEVELOPMENT

CHAPTER 4

PHYSICAL EVALUATION

To begin a training program for any new athletic endeavor, it is beneficial to begin with a physical evaluation. A self-examination will allow you to note your physical tendencies, shortcomings, and attributes. It will help you highlight your strong points and implement changes in the areas needing improvement. Some of the evaluation methods discussed below are subjective: that is, there is room for discussion and interpretation. Other tests are objective, with accurate measurements made to determine a physical tendency or ability. However, even the results of an objective test are open for interpretation in light of each person's other physical and mental characteristics.

Body types. There are several ways to classify bodies according to type. One is the ectomorph, endomorph, and mesomorph classification system. The ectomorph tends to be tall and thin; the endomorph short and thick; and the mesomorph medium and muscular.

The extreme ectomorph may lack strength and coordination for some riding and horse-care related tasks but possesses the physique to be a graceful and attractive rider. The extreme endomorph, on the other hand, may lack the fitness and overall athletic condition necessary for top-notch riding but has the potential to be a very stable, secure rider. The extreme mesomorph may be muscle-bound and as a result may lack flexibility and smoothness but has the capacity to be a top rider in events requiring strength and stamina. Most people fall into an intermediate body type with characteristics from two or three of the categories.

Another way to classify body types, based on the relationship between height, weight, and proportions, is by hormonal influence. Using women as an example, Body Type 1 is dominated by the gonadal gland. Characteristics consist of a flat stomach, small waist, and predominant rear. People in this category often wear much smaller-sized tops than bottoms. Weight is gained in the hips and thighs. The majority of women fall in this category. Type 1 will usually have a low, stable

The endomorph, the mesomorph, and the ectomorph.

center of gravity which is very desirable for riding. If this rider gains too much weight, however, the mobility of her pelvis and thighs might become impaired.

Type 2 is dominated by the thyroid gland and is thought by many to be the ideal female hourglass type. It is characterized by a well-proportioned skeletal structure that gains weight evenly. Type 2 would be an ideal body for a rider.

Type 3 is dominated by the adrenal gland. Characteristics include broad shoulders, undefined waist, and flat buttocks. The upper body is the area that gains weight first. Somewhat more like a masculine figure, Type 3 may tend to get top-heavy, which could cause balance problems in riding.

Type 4 is dominated by the pituitary gland. With childlike features, this type has few curves, gains weight evenly, and has a head that appears large for the body. Type 4 is the only type that may have insufficient stature and strength for some of the physical demands associated with riding.

Although thinking about where you fit in a body type classification system gives you a starting point, the answers to the following questions may be more significant:

■ Are you fat or are you big?

■ Are you skinny or are you thin?

■ Are your muscles strong or are they tense?

■ Are your muscles weak or are they too relaxed?

As you evaluate yourself physically, it is best to compare yourself to a personal ideal instead of to a friend who may be of another structure. Never compare yourself to models in photographs. The models that advertisers use to create selling images are chosen to establish feelings of inadequacy in consumers so that they buy what the model is wearing or using. It is not so much what you've got but how you use it that will determine how well you do in riding.

To become a good rider, you need good overall health. You should have good posture with strong abdominal muscles to support your lower back. You should be at a moderate level of fitness so that you are able to handle the demands for greater oxygen and heat dissipation that active riding will require. Your blood pressure should be normal. As you perform some of the following tests to evaluate your physical potentials for riding, you may need a friend to help measure or to observe your performance. It will also be helpful to have access to a large mirror, a tape measure, and a protractor for some of the tests.

PHYSICAL TRAITS FOR RIDING

Conformation

The shape and proportions of your skeleton will affect how well suited you are for riding. For example, are you knock-kneed or bow-legged? Stand straight with your feet flat on the floor. Bring your legs together and see which parts of your legs touch first. If it is your knees, you are likely to be knock-kneed. If your ankles touch first, you tend to be bow-legged. And if your ankles and knees come together at the same time, you are neither. Many men are slightly bow-legged which is somewhat of an advantage to fitting the normal configuration of the horse's barrel. Extensive riding, however, can perpetuate and exaggerate the bow-legged condition and cause trouble on the inside of the knee joints. Women, on the other hand, tend to be slightly knock-kneed, which may cause stress to the outside of the knee joint and make it difficult to attain "proper" leg position. Riding can have a positive effect on a knock-kneed tendency and may even straighten the legs.

How correct is your leg alignment? Stand on a hard surface with your feet hip width apart and put all of your weight on one foot. Lift the other foot, let the leg hang loosely and use your hands to point the knee straight forward. Then step down on the straightened leg. Now where does the foot point? Check the other leg. The crookedness in your leg or legs may be located in your hip, knee, or ankle and may present a persistent problem if you try to attain an "ideal" leg position while you are riding.

Strength

Strength is the ability of the body or a part of the body to apply a force initiated by a muscle contraction. For each movement in your body, a muscle (or more likely a group of muscles) contracts or shortens. This muscle is called the agonist or prime mover. As this muscle contracts, an opposing muscle (the antagonist) provides some resistance to prevent over-contraction of the agonist or wobbling of joints. Ideally, this occurs in a state of cooperative antagonism: an optimal amount of resistance is applied by the antagonist muscle so that too much drag is not put on the agonist; otherwise, it will not be able to do its job effectively.

Think of the upper arm, for example, with the triceps at the rear and the biceps at the front. To bend your elbow, the biceps contracts, but the triceps must allow this by relaxing. The biceps is the agonist and the triceps is the antagonist.

The roles are reversed when you straighten your arm: the triceps contracts as the biceps relaxes, so the triceps is the agonist and the biceps is the antagonist. In most movements, groups of muscles are involved on each side of this cooperative antagonism.

The muscle strength you use while riding is dictated by an agonist's ability to contract, the antagonist's ability to relax yet provide sufficient stability, and the mechanical relationship of the bones involved in the movement. Most muscles' contractile ability can be increased through strength-training exercises. The ability of muscles to relax affects coordination and timing and can be improved through practice. And although the mechanical action of the bones is relatively fixed, in riding, the net result can sometimes be improved by a change in rider position.

You use two kinds of strength while you are riding: static and dynamic strength. Static strength involves isometric muscle contractions and applies a force at a particular point without going through a motion. Examples of this still strength would be the aids for a half-halt in dressage or a turnaround in reining. In each case the rider performs an isometric contraction of back, pelvis, and legs without a visible movement of the body. In the half-halt, the rider holds the contraction for just a second to heighten the horse's awareness of the aids, then relaxes the contraction. In the turnaround, the contraction may be held for several seconds as the horse performs its spin.

The other kind of strength is dynamic strength, a force applied through a specified range of motion, usually observable. The rider performing a posting trot is using strength in motion or dynamic strength. So is the roper who makes a loop, stands, and throws the rope at a steer, all the time maneuvering her horse.

Riding does not require phenomenal dynamic strength. In fact, although riders should be muscularly fit, they do not need to be body builders. A certain amount of strength is necessary, however, to increase muscular endurance and agility. Very little upper-arm strength is required in riding and, in fact, too much may actually hinder a rider.

Muscles for riding. The muscles called on most often when riding are those of the legs, hips, buttocks, and belly.

The *satorius* is a leg muscle that runs across the front of the leg from the outside of the hip around to the inside of the knee. This is the muscle that allows you to rotate your lower leg inward (toes forward!) so that the side of your calf is against the horse.

The hamstring muscles located at the back of the thigh are comprised of the *semitendinosus*, the *semimembranosus*, and the *biceps femoris*. These muscles rotate the leg inward and outward, tip the pelvis back, and pull the seat bones down into the saddle.

The *quadriceps* are the major muscles at the front of the thighs. They are used mostly by riders who jump, post, or ride with short stirrups. Although the quads are essential for lifting your legs as you walk, for most riding activities they are not called upon as greatly as the hamstring group at the back of the legs.

The *adductors* are located inside the thigh. They allow you to squeeze and grip the saddle to stay on in rough situations. Yet when they are contracted, they immobilize the pelvis and lift you out of the saddle. So, in essence, you lose your deep, following seat when you grip with your adductors.

The *gastrocnemius* muscle is located in your calf area and the Achilles tendon

Muscles important in riding.

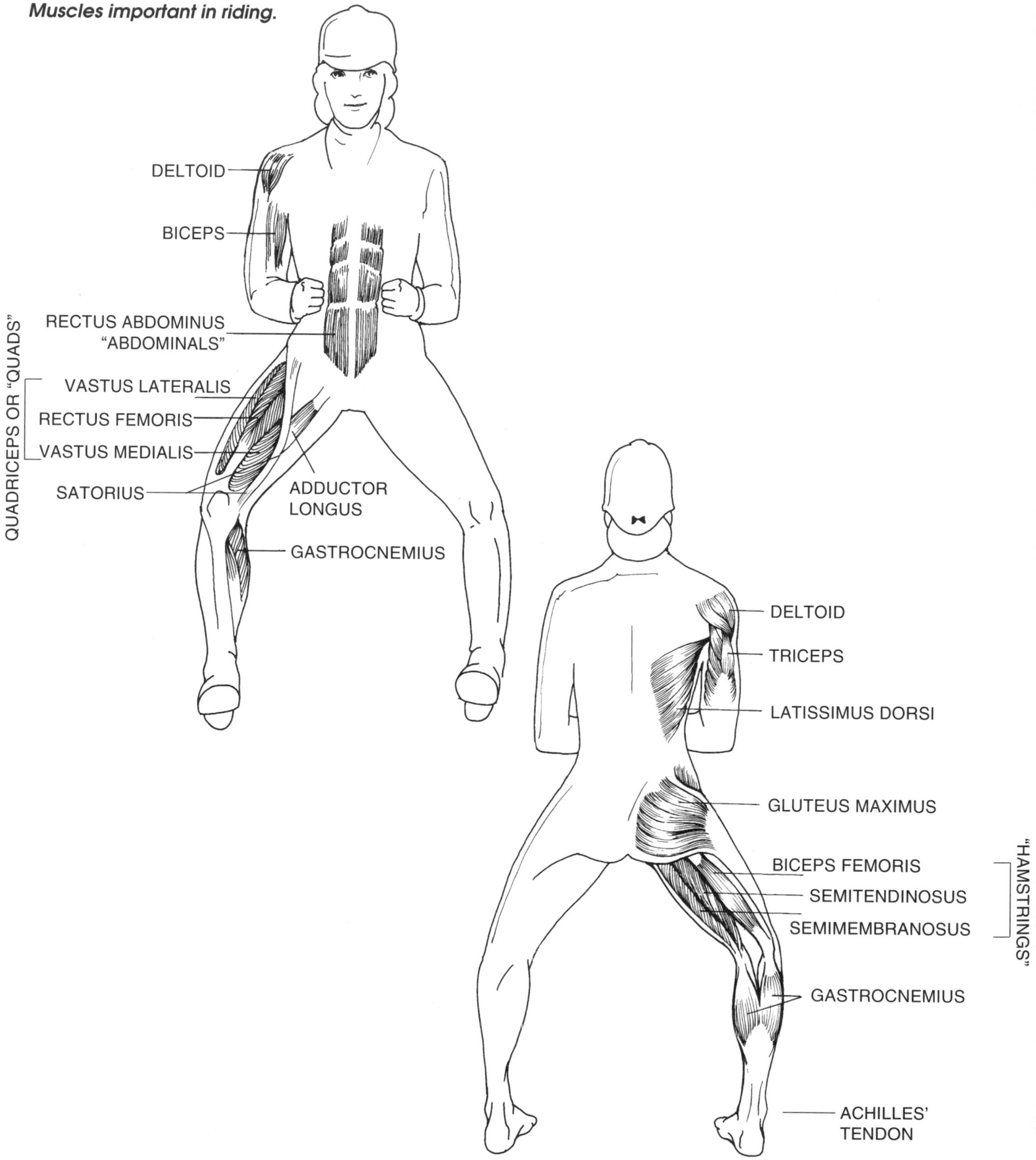

DELTOID

BICEPS

RECTUS ABDOMINUS
"ABDOMINALS"

QUADRICEPS OR "QUADS"

VASTUS LATERALIS

RECTUS FEMORIS

VASTUS MEDIALIS

SATORIUS

ADDUCTOR
LONGUS

GASTROCNEMIUS

DELTOID

TRICEPS

LATISSIMUS DORSI

GLUTEUS MAXIMUS

BICEPS FEMORIS

SEMITENDINOSUS

SEMIMEMBRANOSUS

"HAMSTRINGS"

GASTROCNEMIUS

ACHILLES'
TENDON

attaches the gastrocnemius to your heel. The tendon and muscle work in concert to allow you to lower your heels and deepen your leg.

The *gluteus maximus* and *gluteus medius* allow you to spread and rotate your thighs outward and can prevent the pelvis from tipping forward. If these muscles are used with too strong a contraction, however, they can actually raise your seat bones away from the saddle and make your seat and leg ineffective.

The abdominal musculature consists of the *rectus abdominus*, the *abdominal obliques*, the diaphragm, and the *iliopsoas*. The abdominal group controls your pelvic movements and allows you to flatten or brace your back, as well as to follow the swinging of your horse's back with a swinging lower back. The iliopsoas attaches to the front surfaces of the lumbar vertebrae and tailbone as well as the inner surface of the pelvis. It is capable of pulling the seat bones forward and flattening the lower back. Your abdominal group is your insurance policy against lower back problems that can be associated with improper riding. The more work the abdominal group can take off of your vulnerable lower back, the better. To see how capable you are of flattening your lower back, stand 12 inches from a wall and rest your head, shoulder blades, and buttocks against the wall. Run your hand between the wall and your lower back. Now use your abdominal group to bring your lower back as close to the wall as possible. If it touches the wall, you are well suited for riding. If not, you need to increase the strength of your abdominal group.

To test your abdominal strength, lie on the floor with your knees bent and your toes hooked under a piece of heavy furniture. With your hands behind your neck, do as many bent knee sit-ups as you can in a minute. You should be able to do twenty to thirty.

Balance

Body type, shape, and proportions, as well as your weight, will determine the location of your center of balance. The terms equilibrium, center of balance, center of gravity, and center of mass are all basically the same. They indicate a point in a body around which its weight is evenly distributed or balanced. Your center of balance is located somewhere within your abdomen in the vicinity of your belly button. The lower a rider's center of balance, the closer it will be to the back of the horse. Therefore, if you have long legs, wide hips, and a short, light upper body, your center of balance will be low and you should have a natural physical advantage as a rider. On the other hand, the short-legged, long-waisted individual with a heavy upper body will have a higher center of gravity and may have to work harder to maintain a stable upper body position.

The exact location of your center of gravity will be affected by the location and type of tissue mass and the proportion of your upper body to your lower body. Men tend to have higher centers of gravity because more of their mass is located in their upper bodies; women's lower center of gravity is due to the majority of their weight being located in the lower portion of their bodies.

One way you can tell how your proportions affect your center of gravity when you ride is to measure your upper body measurement (sitting height) and compare it to your lower body measurement. Sit straight on a table and look ahead. Have a friend measure the distance from the table to the top of your head. Subtract this from your overall height to get your lower body measurement. Now divide your lower body measurement into your upper body measurement and you will find their mathematical relationship. If the ratio of your upper body to your lower body (UB:LB) is 1.1:1 or 1.2:1, you tend to have proportions that are an advantage in riding. A high UB:LB ratio (1.3:1 or greater) means a person has very short legs and a tall upper body. The proportions have become so extreme that the person's legs are too short to stabilize her tall upper body easily. The rider with very long legs and a very short upper body has a relatively low UB:LB ratio (less than 1.1:1) and has an advantage in riding because of the added stability provided by the long legs. To illustrate this idea in your mind, think of three people that are all 5 feet 6 inches, or 66 inches tall. Person A's sitting height is 30 inches; B's is 36 inches; C's is 40 inches. Their lower body measurements are 36, 30, and 26 inches, respectively. The ratio UB:LB is .8:1, 1.2:1, and 1.5:1, respectively.

Sense of balance. To determine your innate ability to balance, stand with your hands on your waist and shift your weight to one of your feet. Bend the other leg at the knee and place its sole on the inside of the opposite knee with the toe pointing toward the floor. Close your eyes and see how many seconds you can balance without deviating from the position or opening your eyes. Try the other leg and take the best time out of three. You should be able to balance like a stork for at least 30 seconds.

Sidedness. Somewhere around 90 percent of humans are right-handed but this does not necessarily mean that right-handed people are destined to be right-sided. Many people are born with or can acquire the ability to use both hands fairly

equally except for very specialized tasks such as writing. People involved in balanced physical activities, such as riding, often become ambidextrous. However, some associated horse tasks such as shoveling, raking, and sweeping tend to develop one-sidedness unless practiced from both sides of the body.

Although it is ideal to handle a young horse from both sides from birth, many horses are often handled solely from the near side (the left side of the horse) from birth. This makes a horse tend to bend more easily to the left than to the right. The right side would be termed supple, able to stretch to the left; the left side, on the other hand, may be stiff, sometimes unable to stretch toward the right, making turning right more difficult.

The ambidextrous rider who is equally strong and coordinated on both sides has a better chance of working through a horse's stiffnesses and making the horse more balanced from left to right.

As you ride or watch yourself ride on video, see if you show signs of sidedness. Do you twist your body at the trot or canter? Sometimes at the posting trot, a rider will twist at the waist as she rises. Also, depending which lead the horse is loping on, a rider may shift her hips diagonally in the saddle and let one leg (usually the same one as the horse's leading foreleg) move ahead of the other. Do you routinely lose one of your stirrups? This may indicate that you are contracting one side of your body; as it becomes shorter, you lose contact with the stirrup tread and the stirrup is lost. Do you cock your head to one side? A rakish tilt may be fashionable for modeling a designer hat, but it can also start a dangerous chain of events that prove disastrous for your spine. Measure arms and legs at similar positions and compare measurements. More bulk, greater inches, on one side indicates that it is a dominant side.

Flexibility

Flexibility, or the range of motion of a joint, depends on the bone structure of the joint, the type and amount of tissue surrounding it, and the extensibility of the ligaments, tendons, muscles, and skin that cross over it. Inactivity can cause a rider's muscle and connective tissues to lose their extensibility. The flexible rider moves smoothly. A lack of flexibility can result in improper movement, poor form, and injury. Too much flexibility, however, can also cause injuries such as dislocations and sprains.

A rider must be especially flexible in the pelvis and hips. To increase your flexibility in an area, use slow stretching exercises, not bouncing exercises. Bouncing can cause you to dangerously exceed a tissue's extensibility.

The rhythmic movement of the horse can improve the rider's flexibility because the movement of the horse closely approximates the movement of the human pelvis during walking. That is the basis of hippotherapy, a form of physical therapy that uses a well-trained and balanced horse to improve a person's posture, balance, muscle tone, mobility, and function. A rider receives about a hundred vibrational impulses per minute through the horse's undulating back. These movements are transferred to her pelvis, spine, and shoulders. As her body moves with the horse, she can begin to assimilate an improved posture.

A hippotherapy horse must have had training up to the equivalent of second- or third-level dressage in order to move with balanced, even gaits. A stiff or imbalanced horse that moves with an irregular rhythm would not be helpful and could cause further problems. (See Chapter 11, "Your Mentors," which discusses the importance of using an appropriate school horse for the development of any rider.)

Hip flexibility will allow the rider to attain a comfortable, effective seat and to follow the horse's movements. To get an idea of how flexible your hip joints are, lie on your back with your head and hands on the floor. With one leg stretched out in front of you, bend your other leg at the knee and bring it close to your chest. Have a friend note the angle between your spine and femur. (When your knee is pointing toward the ceiling your femur is at a 90-degree angle with your spine and the floor). You should be able to close the angle to at least 60 degrees. Check to see if one hip is more flexible than the other.

Thigh muscle suppleness is important for enveloping the horse's barrel and for developing an independent use of the legs to aid the horse. Sit on the floor with your back straight and one leg straight out and parallel to a wall. Move the other leg away from it making as wide an angle as you can. If you can only open your legs 90 degrees or less, you need stretching exercises to limber up for riding.

To test your thighs further, bend your knees and bring your soles together. Move your feet in as close to your crotch as you can, keeping your knees as close to the floor as possible. If the distance between the bottom of your knee and the floor is more than 9 inches, you need to do some suppling exercise. Note if one of your knees is higher than the other.

The "heels down" position desired in many forms of riding requires that your *hamstrings, gastrocnemius muscle,* and *Achilles tendons* ("heel cord") are stretchable. Sitting on a chair with your legs

Flexibility adds to a rider's performance.

stand with your arms in front of you, elbows and wrists straight. Hold a rope in front of you, then without bending your elbows bring the rope up over your head and behind you, letting it slide through your hands enough to let you bring your hands behind your back. Can you bring your arms behind you at all? Which shoulder limits you? How far apart do you need to hold your hands in order to do this? If you are between twenty-five and forty-five years of age and can keep your hands closer than 35 inches apart, you're looser than average. If they must be 45 inches or more apart, you're tighter than average. If you are extremely loose, you may be prone to shoulder dislocations, so you should strengthen your upper body.

straight out in front of you, flex your ankle so that your toes reach backward as far as possible toward your shins. If the angle your sole makes with the back of your calf is 80 degrees or greater, you need to do exercises to stretch your calf muscles and Achilles tendons.

Also, in order to achieve the long, low leg desired in western and dressage riding, your *hamstring muscles* and those of your *lower back* should be loose. Stand with your knees straight and your feet flat on the floor, hip width apart, and bend at your waist to reach for the floor. Do not bounce, as this is dangerous and will result in an inaccurate indication of your stretchability. This exercise will probably give you a burning sensation along the back of your legs. The tips of your middle fingers should at least touch the floor.

Shoulder flexibility is determined by joint configuration, muscle bulk, amount of fat around the joint, and the ability of the ligaments, tendons, muscles, and skin around the joint to lengthen and shorten. To test the flexibility of your shoulders,

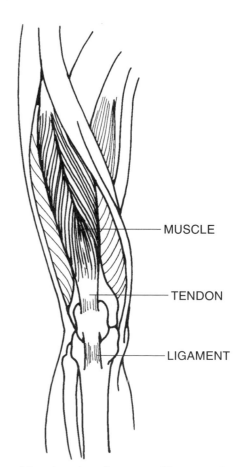

MUSCLE

TENDON

LIGAMENT

Muscles, tendons, and ligaments.

The thigh must be supple so that it fits the horse's barrel.

If you are tight, you need to stretch and supple this joint so that you don't develop tendinitis.

To check which shoulder is your limiting factor, stand with one arm behind your back reaching up toward your shoulder blades. Bring your other arm over the top, reaching toward the other hand. Your fingertips should at least be able to touch. On which side are you most comfortable and most able to perform this exercise? You may find that one shoulder is limited in its reach both from the upper and lower positions.

Agility. Maneuverability, or your ability to change your body direction and parts rapidly and smoothly, is essential for jumping, reining, roping, cutting, and upper-level dressage. It requires flexibility, timing, and coordination.

Timing. Timing is the quickness with which you react and respond to a stimulus or a situation. The more advanced you become in your riding, the more critical your timing becomes. The more accurately and quickly you can respond to a horse's movements, the better able you will be to keep the horse in proper behavior and form. Each time a horse makes a movement, you should encourage or discourage him from repeating that behavior. You do that using your natural aids: your mind, seat, upper body, legs, and hands.

When things go well, the sequence goes as follows:

- you prepare the horse
- the horse is listening
- you apply the aids
- the horse responds properly
- you acknowledge the proper response by slightly releasing the aids
- the horse continues in a more relaxed manner
- you follow the movement of your horse
- the line of communication remains open between horse and rider for the next maneuver

When there is a communication problem between rider and horse, the sequence might go as follows:

- you prepare the horse
- the horse is not really listening
- you apply the aids anyway
- the horse responds improperly
- you correct the horse and bring it back to its original movement
- the horse responds but is tense
- you re-apply the aids for the desired movement with more force this time

Timing is necessary for the application as well as the release of the aids.

- the horse responds properly this time but is increasingly tense
- you continue applying the aids to "hold" the horse in the proper movement
- the horse stiffens
- you stiffen
- the quality of the work deteriorates
- the lines of communication have been closed between horse and rider.

To avoid confusion, deliver the correct response quickly. Your *reaction time* is the time between a stimulus (oops, you feel the horse has taken the wrong lead) and the initiation of your response (the coordination of your back, seat, legs, and hands bringing the horse back down to the trot to try again).

Your *reflex time* is when there is no thought process between the stimulus and the initiation of your movement. You do not have to consciously think or remember what to do next; you do not have to figure out or decide *why* it is the best thing to do; you do not have to review how to do it. A reflex is essentially a shortened reaction time. Your goal as a rider is to develop a set of conditioned reflexes to be used when you are riding.

Some of your innate reflexes, those with which you were born, need to be overridden, so to speak. For example, the reflex of a green rider to a horse moving fast is to grab with the legs and pull on the reins. This makes the horse's engine go faster and the horse stiffens so he goes even more quickly and choppily. Instead you will need to learn a set of conditioned reflexes that allow you to deepen your seat in order to burden the horse's hindquarters, slow his gait, and relax him. This will take time for you to develop.

At first, you will feel the horse do something, then you will use your brain to decide what to do and how to do it, and then you will do it. Eventually, you want to have a series of feeling/action reflexes — when you feel a certain thing, your body automatically takes a particular action. Those riders who are called "naturals" are either those whose innate reflexes seem to be automatically appropriate or those who can quickly develop proper reflexes.

Shortening your reaction time is important so that you can catch a horse's mistakes quickly and correct things before they get out of hand. If you miss the opportunity, you may have to wait until the next stride or the next time around the course, or you may even need to bring the horse down to a walk and start again from square one.

Another aspect of your reaction time will become obvious if you compete in gymkhanas, races, or roping events. Your reaction, and that of your horse, to buzzers, whistles, shots, or flags will have to be honed by practice.

In fact, practice is the most valuable way to improve your sense of timing, providing you don't practice something so many times that you make your horse hyper-anticipatory, resentful, or sour. Imagine or anticipate the action just prior to a particular movement in order to get your muscles ready. Preparatory signals are useful when developing timing. Although we don't often use "Ready, set, trot" in riding, a verbal command from your instructor, such as "Prepare for a trot," can be a helpful preparatory cue. When practicing alone you can use quiet verbal or mental preparatory commands to help develop a sense of timing for the aids.

To be sure your sense of timing is functioning at its best, pay attention to your diet during the hours immediately preceding your ride, whether it is a competition, a lesson, or practice. (See Chapter 7 for more details.) Also, be sure not to be too tired when you ride, as fatigue greatly increases reaction time.

Coordination

Your ability to perform a series of specific movements correctly, effectively, smoothly, and gracefully reflects your neuromuscular skill level. Since your nervous system is the last to respond to the effect of learning a new skill (your heart, lungs, muscles, and connective tissues are conditioned early in your training), it may take years to fine-tune your neuromuscular skills. That's one of the reasons why functional fluidity is so greatly admired in top-notch riders — it takes so very long to develop!

Neuromuscular skill is the co-ordination of what muscles to use, when, and how intensely. The proper muscles act in response to nerve impulses that are developed by practicing the various components of riding: sitting the collected canter, running down for a sliding stop, initiating a lope depart, posting the trot, and assuming the two-point position, to name just a few. Be absolutely sure that you are practicing a component correctly because it will become a habit whether it is right or wrong. If you ride incorrectly, you may be faced with a very difficult and time-intensive relearning process. Many riders have lamented, "I rode wrong for twenty years and am now trying to retrain my body to ride correctly." It is much more difficult to change deeply ingrained old habits than it is to learn correct ones the first time around.

Various factors will affect your ability to hone your riding skills. One is your *proprioceptive sense*, or your awareness of body position transmitted by your vision, the balance organs in your inner ear, and pressure receptors located all over your body surface and in your joints.

Your *vestibular reception* takes place in organs located in the non-auditory portion of your inner ear. The semicircular canals there are filled with fluid. When this fluid moves in response to acceleration, deceleration, and rotation, it pushes hairlike *cristae*. The cristae's movement is then transferred to the brain as body position information. The *utricle* and *saccule*, also located in the inner ear, are involved in registering information on body position.

People with vestibular reception problems may have a learning disability when it comes to some aspects of riding. One possible cause of dyslexia is an impairment of the vestibular apparatus. Symptoms of a dyslexic person may include balance and coordination dysfunctions, inability to concentrate, poor sense of direction, fear of heights, dizziness and motion sickness, and feelings of inferiority and clumsiness. It is clear that these characteristics could cause problems, though not insurmountable ones, in learning how to ride.

Your *kinesthetic reception* is based on information about your body parts and position that your brain receives, without assistance from your vision, from receptors located throughout your body. To demonstrate, close your eyes and hold your arms straight out from your sides, parallel to the earth. Now bring your palms together in front of your face. Feel how accurately you were able to join your palms. Now do it again. Did your accuracy improve? Now try to touch the tips of your index fingers together and observe. This ability to use your body without vision allows you to use your hands effectively on the reins while you are looking forward between your horse's ears at the upcoming terrain.

Kinesthesis is made possible by receptors located in muscles, tendons, and joints. The receptors in muscle spindles are sensitive to changes in the length and the tension of the muscle fibers. They are responsible for the initiation of the stretch reflex — the automatic urge of the body to oppose a strong muscle contraction with a stretch. Receptors located in tendons are sensitive to high tension located there and they create a protective mechanism for dangerous hyperextension of joints and muscles. Other receptors, located near joints, send signals to the brain about the overall position of the body.

Your ability to control the impulses your brain dispatches to your antagonists will result in smooth and efficient movement. Fatigue will also be postponed because the prime movers won't be working as hard as they would if the antagonist muscles were resisting too much. The main way you regulate nerve impulses is through mental and physical relaxation. This will allow you to control physical anxiety and emotional surges. Relaxation can be achieved through centering and visualization techniques covered in a previous chapter.

Durability

Durability is the toughness, strength, and soundness of your joints. To test the durability of your knees, perform the following tests and note any difficulties you have. First run in place evenly noting the sound and feel of each of your feet hitting the ground and the symmetry of your entire body as you run. Then hop on one leg for 15 seconds, then the other leg. Alternate back and forth for several minutes. Note if there is a change in your balance or in the smoothness of your joint movements. Now do five or six full squats. Are you able to sit down on your ankles? Finally, walk across the floor in a full squat position. The strength and correctness in your ankles and knees, two common areas of rider problems, will be demonstrated by your ability to perform these tests symmetrically and without pain.

Fitness

Endurance is resistance to fatigue and the ability to recover quickly from fatigue. There is a distinction between muscular endurance and cardiopulmonary (CP) endurance. Your level of muscular endurance will be painfully evident in your leg and back muscles after a long ride. Your level of CP endurance may be revealed after only five to ten minutes of posting trot work. In order to increase either, you must work beyond your present level of endurance to experience the effect of progressive overloading. Although long continuous work at the same pace does show benefits, greater benefits are received by using interval training.

Interval training consists of brief work periods or "works" interspersed with rest or light work. You might begin with ten minutes of trotting followed by a five-minute walk break, repeating for three-quarters of an hour. Within two months (providing your horse's condition will allow) you may have moved up to an hour-and-a-half session with fifteen-minute lope and trot works and two-minute breaks. With interval training you can increase your endurance potential by:

- increasing the number of works
- increasing the length of the works

- increasing the intensity of the works
- decreasing the number of rest periods
- decreasing the length of the rest periods
- performing the work in hot weather

Breathing. As you ride your horse, your lungs take in oxygen and expel carbon dioxide, just as your horse's do. Your resting respiration rate will probably be in the vicinity of twelve breaths per minute. During heavy exercise, it will be in the range of thirty-five to forty-five breaths per minute. Even when you are exercising at your maximum, your lung capacity is usually adequate. In fact, during moderate exercise, many people take in four to five times as much oxygen as is actually consumed by the muscles.

It is usually not the lungs that can't keep up, but the heart. The heart's job is to pump oxygen-rich blood to the muscles and return blood loaded with carbon dioxide to the lungs to be purified — in essence, aerobic metabolism. If you are unaccustomed to strenuous exercise, your blood may not be able to deliver enough oxygen to your muscles. Your muscle cells solve the problem by kicking into anaerobic metabolism. They burn stored fuel (glycogen) incompletely, using little or no oxygen. Although anaerobic metabolism continues to fuel the muscles, the incomplete combustion results in an increased build-up of the toxic by-product lactic acid. Lactic acid is responsible for muscle fatigue and the "muscle burn" of overexertion. As your body neutralizes and dispels the lactic acid, carbon dioxide is produced. You breathe faster to expel the carbon dioxide and this can make you feel "short of breath."

Gradually there will be a shift back to

aerobic metabolism when the muscle fuels (carbohydrates) have been restocked and are once again being burned in the presence of oxygen. When you reach this point, you may feel you have gotten a "second wind."

As you become more conditioned to exercise, you increase your aerobic potential and thus need to dip into your anaerobic stores much later in the exercise period. Your muscle tissues are subjected to less lactic acid. You are able to rid your body of lactic acid more efficiently and return to the "clean burning" aerobic state sooner. The capacity of your heart and blood vessels to deliver oxygen-rich blood to the muscles increases, along with the ability of your lungs to remove carbon dioxide. You can ride more intensely, therefore, without becoming short of breath. The more regularly — not necessarily the more vigorously — you actively ride, the more efficient your breathing becomes. You will begin to take fewer, deeper breaths.

THE PHYSIOLOGICAL DIFFERENCES BETWEEN MEN AND WOMEN

Because there are definite differences in physical traits between men and women, they have distinct advantages and disadvantages as they learn to ride. These are general tendencies and do not hold true in all cases.

Women are innately more flexible and loose-jointed, which allows them to follow the movements of the horse more smoothly, but it also means they can be more easily unseated and may be more likely to suffer joint injuries. Women also have greater manual skills and dexterity

which may allow them to be more talented and subtle with rein aids. The tightness of men's joints, muscles, and skin may make them less responsive to the movements of the horse, but it does allow them to hold a more correct position more easily.

Women tend to carry 20 percent of their weight or more as fat, which provides greater cushion, insulation, and tolerance for cold. It can, however, make the body difficult to cool in hot weather, may inhibit range of motion, and may undermine level of fitness. In addition, a woman's thermostat or sweating threshold is higher, so a woman must be hotter before she begins to sweat in order to cool down her body temperature. Men's lower fat stores may allow them to become more fit but their lack of padding and insulation may make riding less comfortable for them, especially in cold weather. Since men begin to sweat at a lower temperature they begin cooling their bodies sooner.

Women have lighter bones and less durable tendons and ligaments, which results in a lighter load for the horse to carry but can result in injury, especially to the knees. The only disadvantage of men's heavier framework is that it is a heavier load for a horse to carry in speed or endurance events.

Women are generally 3 to 4 inches shorter and 25 to 30 pounds lighter than men so they can often ride smaller horses. However, men's generally longer legs often fit around a horse's barrel better than a woman's do.

Women are bottom-heavy with a low center of balance. The heaviest area of their bodies is between the hip bones. This creates more resistance to the movement of the lower limbs, resulting in a better leg balance for the advanced rider, but may cause the beginning rider diffi-

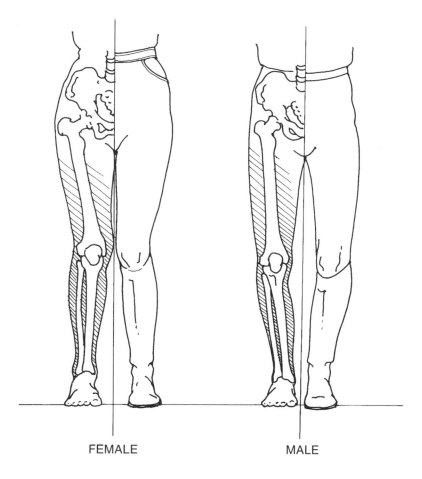

FEMALE   MALE

culty as she tries to move her legs into various positions. Men are often top-heavy with a higher center of balance located somewhere around the waist. This makes the lower limbs more movable, which is both an advantage and disadvantage to the beginner rider: leg adjustments come more easily but leg stability is more difficult to maintain. Top-heaviness combined with men's higher center of gravity may make a male rider become more easily unbalanced.

Women tend to be less strong and have a lower capability for building muscles. To some extent this is a limiting factor but since extraordinary strength is neither necessary or desirable for riding, the female rider will likely not become muscle-bound. More importantly, she will resort to training by psychology rather than strength. A man's greater strength

can tend to make him overpower a horse, especially with upper-body strength applied through the reins.

Saddles have traditionally been designed for male riders (for cavalry, ranch work, and European sports), yet the majority of today's riders are female. The fundamental anatomical structure of the female pelvis and legs may make some facets of proper riding more difficult for a female, and an inappropriate saddle will exaggerate the tendencies. Women have a wider pelvis, more widely set seat bones that angle outward, hip sockets that face outward, and a tail bone that tips strongly out behind the spine. Men tend to have a narrower pelvis, nearly parallel seat bones, hip sockets that face forward, and a tail bone that is more vertical. These characteristics cause a woman's legs to be oblique toward the knee (a tendency toward being knock-kneed). This may make it harder for a woman to relax her thighs and let them hang down along the horse's side; instead they may point outward at the knee and tend to come forward and upward when riding. In addition, with the tail bone located behind the lumbar vertebrae, women tend to have a naturally hollow lower back. This requires more muscular control and precise positioning to bring the seat bones under, tuck the tail bone, and achieve a flat lower back. Female riders tend to tip forward on the pubic bone, ahead of the center of balance. This is especially common in young female riders. Improper instruction can cause a female rider to exaggerate the hollow lower back and may set the stage for chronic back pain.

The male's narrow pelvis and straight legs (sometimes with a tendency toward bow-leggedness) allow the thighs to hang down along the horse's barrel. His pelvis is more upright, resulting in a flatter lower back and a more naturally stable position. The male's near-parallel seat bones and near-vertical tail bone are perfectly suited for a deep, following seat. The seat bones are able to rock freely backward and forward in contrast to the muscular effort required for a female rider to do the same. The naturally tucked tail bone of the male rider allows him to sit down more effortlessly on a trotting or cantering horse than a female rider who must constantly exert muscular energy to tuck the tail bone and flatten the lower back. The narrower pelvis of the male

rider does not seem to present difficulties when riding a wide horse.

The previous generalizations related to male and female anatomy will vary in degree according to the individual. Some women have a more masculine pelvic configuration while some men show female characteristics in that region. Feedback from a qualified instructor regarding pelvic and lower back action is essential to safeguard the health of your spine.

Fitness is important for the active rider.

CHAPTER 5

EXERCISE

The physical development of the rider takes place in stages. You will probably not progress through the phases in a strictly linear fashion, however — you will find yourself working on elements from several phases at once. Most riders discover the importance of the following stages in the approximate order listed. As with other sports, however, every person needs to design her own personalized development program.

PHASE I requires that you have enough muscular strength to perform riding basics for about one hour. You can improve your strength by riding, by doing isometric exercises and/or isotonic exercises on a regular basis, or by participating in "companion" sports or activities such as cross-country skiing.

PHASE II requires that you begin increasing your muscle and cardio-pulmonary endurance. Use interval training principles to help you achieve the maximum effect. For example, ride posting trot for ten minutes, then walk for five minutes and repeat the cycle for as long

as you can perform well — up to an hour or more. Such a workout will later be the basis for your more vigorous work sessions.

PHASE III requires that you increase your flexibility so that you can begin working on a more correct riding position and a more effective use of the aids. You can use isotonic stretching exercises that lengthen your upper body, stretch your legs, or increase the range of motion of your hips. Beware of some isotonic exercise programs that advocate many repetitions — you may develop too much muscle bulk, which may hamper your flexibility.

PHASE IV consists of fine-tuning your weight, posture, and muscle tone. By this time you will have identified trouble spots and can design a diet and exercise program and effect changes in your everyday habits that will help you minimize your problems.

PHASE V involves improvement of your timing. Here you are working to develop an advanced level of skill. Timing is improved by practice and enhanced by strength. Strength allows muscles to contract quickly and with optimal force. Your reaction time is adversely affected by fatigue, drugs, alcohol, altitude, age and, especially, lack of practice. When you choose and implement wise habits you stack the deck in your favor.

Exercise

Exercise has many benefits. It improves muscle tone, posture, and balance; it creates a sense of satisfaction and well-being; and it burns excess calories to produce energy. Active riding does burn calories. As you sit and read this book, you are using approximately 30 calories per hour. If you were riding your horse at a walk, you would use 120 to 240 calories per hour; at a trot, 240 to 420 calories per hour; at a canter, up to 480 calories per hour.

As you exercise, your body is learning how to produce energy more efficiently, cool itself, and minimize the production of waste products. All of this means a healthier body and a happier state of mind. One characteristic of any athletic pursuit, and certainly of riding, is that you will receive the most exercise benefit when you and/or your horse are not in shape. During those times, you have to work harder to stay in position and keep your horse balanced and moving forward. Once you are in shape and your horse is well trained and conditioned, most types of riding will tend to give you less exercise benefit.

Breathing

No matter what type of exercise you are doing, you must learn to breathe properly. How much you should concentrate on your breathing depends on the breathing habits you wish to change. First work on breathing exercises when you are not riding. If you concentrate too much on your breathing while you are riding, it might throw off your natural rhythm. Be sure to check periodically as you ride that you actually are breathing. Sometimes when a rider is concentrating on the proper application of the aids or on controlling the horse, she will forget to breathe. She will arrest her breath until an absolute need for oxygen is created, then she will gulp down a huge breath.

It is healthiest to inhale regularly through your nose and exhale through your mouth. Air taken in through the nose can be filtered, warmed, and humidified before it reaches the lungs. This is especially important in a dusty arena. When riding in very cold weather, you may wish to wear a lightweight scarf or ski mask over your nose to help warm the incoming air.

Since your lungs have no muscles of

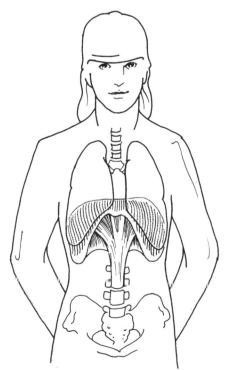

The location of the diaphragm.

their own, they are squeezed by the muscles of the diaphragm (a dome-shaped muscle below the rib cage) and the muscles between the ribs. When you take air in, the diaphragm contracts and flattens, the rib cage is pulled upward to enlarge the chest cavity, and the lungs expand. During exhalation, the rib muscles and the diaphragm relax and return to their original positions, forcing air out of the lungs.

Maybe you have been a "fashion breather" up to this point, sucking in your abdomen to flatten your stomach, filling your lungs until your shoulders are taut and square, and then exhaling by collapsing your lungs and letting your abdomen fall slack. There is another way to breathe but you will have to be willing to let your tummy bulge momentarily.

Close your eyes. Take air in through your nose and send it downward as if to fill your abdomen, not upward to fill your lungs. This allows the diaphragm to follow its natural tendency to expand your abdomen and fill your lungs indirectly and without tension. When you are ready to exhale, do so through your mouth to empty the lungs and "deflate" the abdomen. Wait after exhalation and allow your next breath to arise spontaneously. During this interim, sometimes called the "creative pause," you should feel a welcome emptiness, an unencumbered space into which images, but not internal dialogue, can flow. The pause created by proper breathing rhythms allows for restoration of energy stores. After the pause, breathing resumes with a stronger connection and a greater energy flow.

High altitudes. Because the density of oxygen molecules in the air is lower at higher altitudes, it is more difficult to breathe at altitudes above 6,000 feet. Even if you are in good shape, your body loses

Because of the lower density of oxygen at high elevations, riding at altitudes over 6,000 feet requires conditioning.

3 percent of its capacity to consume oxygen for every 1,000 feet above 5,000 feet above sea level. To perform as well at high altitudes as you do at sea level, therefore, you will have to take in more breaths per minute in order to provide your body with the same amount of oxygen. In addition, since many high-altitude locations are also arid, you will be losing more body water through respiration than you would in moderate humidity, especially taking into consideration the rapid breathing characteristic of high-altitude riding. If you will be riding at altitudes above 6,000 feet, be in good condition, allow for a day of acclimatization before exertion, drink plenty of water, and don't overdo.

Pollution. Avoid riding near heavy traffic and other air pollutants. Since you breathe faster and more deeply when you are riding, you would be taking in more carbon monoxide, sulfur dioxide, and ozone, smog's major component. Ozone levels tend to be the highest during the early afternoon, so avoid riding during

While mounted, close your eyes and tune in to your breathing.

that time. Repeatedly taking pollutants into the lungs can cause coughing, throat irritation, and difficulty in breathing.

Isometric Exercises

Isometric exercises consist of muscular contractions performed in a fixed, non-moving fashion. They are therefore a convenient type of exercise that can be performed in almost any place for short periods of time, wearing everyday street clothing. You can perform isometrics as you drive your car, work at a desk, or wash dishes. An observer will probably not detect that you are exercising.

Breathing is especially important during isometrics or blood pressure can rise, decreasing the flow of blood to your heart. On the other hand, if you breathe excessively (hyperventilate) before you exercise, and then hold your breath, you may faint.

As part of body awareness and strengthening, isometrics help you locate and isolate specific muscle groups and strengthen them through prolonged contractions. The abdominal muscles, for example, which keep the lower back and buttocks deep in the saddle and drive a horse forward, need to be sufficiently strong. The abdominal contraction required for riding might be somewhat contrary to the common perception, just like the misconception regarding breathing. Rather than sucking the abdominal muscles in to achieve a contraction, it is beneficial to learn to push on them without hollowing your back. Place your hands on your abdomen and press your muscles against your palms. To give your contraction added power, exhale as your press your abdominals out and inhale as you relax the contraction. Once you have identified the feeling of an abdominal contraction using your hands, you will be able to perform this isometric exercise anywhere, any time.

To strengthen the inner and outer thighs, find an immovable object (the wall, the side of a desk, a footstool, etc.) that you can place your knees or ankles alongside. Then push outward to strengthen the outer thigh muscles and push inward to strengthen the inner thigh muscles.

Isotonic Exercises

When you have time to change into clothing appropriate for a vigorous workout, you can run through your personalized set of isotonic exercises to strengthen and stretch your muscles for riding. Isotonics are active exercises in which muscle fibers actually shorten to cause movement of a body part. They are exercises in motion, the type to which you have most likely been accustomed all of your life. Here are some of my favorites:

The quadriceps stretch. To improve balance and flexibility of the large muscles on the front of the thigh. If these muscles are tense, they may interfere with the use of the pelvis and lower leg when you ride. Stand on one leg (you may need to grasp a support) and grab your other ankle with the hand on the same side. Smoothly pull your heel toward your buttocks. Keep your back straight and extend your hip (downward). Hold. Repeat with the other leg. You can do this exercise before you mount, during some of your breaks (dismount, of course), and after you ride.

Hamstring stretch. To lengthen the large muscles at the back of the thigh for a deep seat and long leg. Stand keeping one leg straight. Bend the other leg at the knee and move its foot around the front to the outside of the other foot. Bend at the waist and reach for the floor. You will feel the "burn" at the back of your straight leg.

Side stretch. To elongate the side of your body, especially beneficial for a rider with a collapsed side. With your feet hip-width apart, raise one hand over your head. Reach for the ceiling as you stand on your tiptoes and feel your entire side elongate.

Lunge. To strengthen the quadriceps and stretch the gastrocnemius and Achilles tendon. Place one foot 2 feet ahead of the other. Bend the knee of the front leg, keeping the back leg straight and the back heel on the floor. Hold your arms out to your sides, horizontal to the floor, keeping your back straight. You should feel a strengthening in the quadriceps of the front leg and a stretch of the gastrocnemius and Achilles tendon of the back leg. Repeat with other leg.

The stirrup stretch. To improve the balance, coordination, and stretching required for mounting. With one foot flat

Quadriceps stretch.

Hamstring stretch.

Side stretch.

Lunge.

Symmetry Stance.

Stirrup stretch.

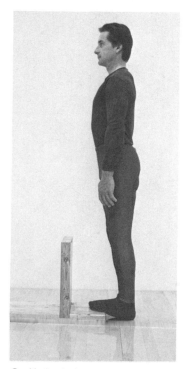

Calf stretch.

Lower back relaxer.

on the floor and the opposite hand on the hips, raise the other leg with the knee bent so the thigh is at least horizontal. Reach the opposite elbow toward the knee. Hold. Repeat with the other leg.

Symmetry stance. To improve your overall balance, symmetry and poise. With your feet placed wider than your shoulders to approximate the position on a horse, arms out horizontally from shoulders, tuck your buttocks as if to sit. Keeping your lower back straight, squat as far as you can toward the floor while keeping your correct position. Regulate your breathing.

Calf stretch. To stretch your gastrocnemius and Achilles tendon in order to help you ride with a long leg and low heel. Stand with your knees straight and the balls of your feet on the edge of a 2 4 inch step. Let your heels stretch down. Hold five seconds. Rest or raise above the step on your toes.

Lower back relaxer. To stretch and relax your lower back and hamstrings and round the lower back; more convenient to perform "on site" than the back stretch below. You can use this exercise before, during, or after riding. With your feet flat on the ground, squat so you are sitting on your heels. Clasp your arms around your legs, rest your chin on your knees, and let your muscles relax. Once you have practiced this relaxer, you will find your body remembers it and will automatically configure in that position when you squat to put bandages or boots on your horse's legs or to clip his legs.

Back stretch. To stretch your lower back and hamstrings and discourage a hollow back (lumbar lordosis). Lie flat on a floor or mat. Bring one or both knees to your chest. Clasp your hands around your upper shins and hug your legs toward your chest for a better stretch. Keeping your back flat, slowly raise your head and touch your nose to your knees. Hold for five seconds. Slowly uncurl. Repeat. Don't forget to breathe.

Abdominal strengthener. To tighten and strengthen the abdominal muscles; also called the knee-bent sit-up. With your knees bent and feet flat on floor, and with your back, shoulders, and head flat on the floor, point your arms forward toward your knees. Exhale and slowly begin lifting your head, one vertebra at a time, to raise your shoulders off the floor. Inhale as you let yourself down just as slowly.

Lateral leg lifts. To improve the range of motion of the hip joint and to strengthen the thigh muscles.

Version A. Lying on one side, support your upper body with a bent elbow. Keeping your lower leg on the floor, raise the other leg in a scissor-like motion, alternating the pointing of your heel and toe.

Version B. In the same position, bend your upper leg at the knee and place the

Back stretch.

Abdominal strengthener.

Leg lift, Version A.

Leg lift, Version B.

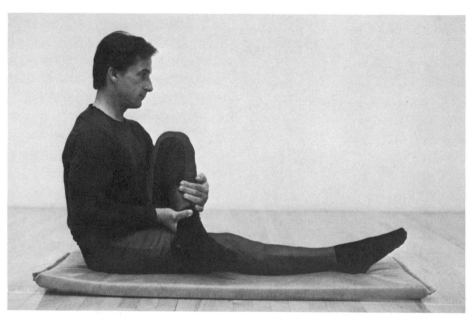

Hip stretch.

foot on the lower leg at the knee. Raise and lower the lower leg. The added weight of the upper leg creates more work for your lower leg.

Hip stretch. To stretch and relax some of those difficult-to-reach hip and buttock muscles that can get tight as a result of riding. Sit on the floor with your legs in front of you. Bend one leg at the knee and cross that foot over to the outside of the opposite thigh. Draw the leg close to your body, keeping your back straight.

Companion Sports and Activities for Riders

So that you can keep in shape year round and in all kinds of weather, find some companion activities that will keep you ready to spend time in the saddle. Staying active year round will minimize weight gain, help prevent muscle atrophy, and reduce chance of injury when you return to active riding. Even if you ride all year, it is a good idea to participate in other sports or activities to make you a well-rounded person mentally and physically.

Bicycling is a good riding-season companion sport, providing the opportunity for great cardiopulmonary improvement as well as fine-tuning your equilibrium. Cross-country skiing is a great "off-season" activity to keep both your muscles and your cardiopulmonary system in shape. The characteristic crouch of the skier uses many of the same muscles as riding does, and the aerobic exercise is unequaled in winter sports.

Indoor cross-training can include various types of dancing. Ballet will improve your flexibility and your ability to execute patterned movements. Ballroom or country and western dancing will improve your coordination and your ability to perform sequenced movements. Fencing utilizes some of the same leg muscles as riding and can improve your reaction time as well as developing a sense of poise in your body carriage. Gymnastics can improve balance, strength, poise, and focus.

CHAPTER 6

BODY AWARENESS

The physical aspects of becoming a rider can be categorized into two main phases:

- learning to feel what your body and the horse's body are doing during motion; and

- learning to influence the way the horse is moving through the way you use your body.

A rider must go through these phases in the above order because it would be ineffective to try to influence a horse (in essence, train him) before first becoming *aware* of your body and the horse's movement. In order to get the most out of your personalized approach to riding, you should consciously free yourself of previously ingrained body sensations and habits and remain open to new body awareness.

LEARNING TO FEEL

The goal of riding is to move as if one with the horse, almost like the mythological centaur. To do this, you must tune into your sense of feeling and tune out extraneous visual and audible clutter. You might be able to do this more quickly by closing your eyes while you are on your horse. Ask a friend to lead you around on a horse as you concentrate on the sensation of the moving animal. If you have a quiet, trustworthy horse and a round pen, you can do this on your own. As you ride with your eyes closed, you will become aware of your center of balance on the horse and find that as long as the horse moves consistently and predictably, your center of balance stays relatively still and you feel OK. When there is a change, a shift to the left or right or backward or forward, you will likely find it disturbing and will want to open your eyes. Don't be surprised if your friend finds this amusing because the giant, threatening movement you perceived might have just been the horse shaking a fly off its neck. Sometimes it is advantageous to exaggerate body positions in order to find your center of balance.

The Rhythm of the Hind Legs

Once you have become aware of your center of balance and how it remains steady or shifts with the horse's motion, focus on finding the rhythm of the horse's leg movements. At the walk, this will be a four-beat rhythm. Because the walk is slow, you will have plenty of time to sense the individual movement of each leg, especially the hind legs where the moving power is generated.

As the right hind leg gets ready to reach forward under the horse's belly to take a step, it and the right side of the horse's back will rise, lifting the right side of your seat and, if your leg is very relaxed, swinging your right leg slightly forward. As the leg lands, the right side of your seat will lower and your leg will tend to move backward to return to its central position.

Once you have become one with the horse at the walk, proceed to the two-beat trot and the three-beat canter. Here it will be best if your horse is being longed by an instructor.

It is important that you take the time to develop a very certain feeling of the rhythm of the hind legs at all gaits. Once you are able to follow the horse's natural rhythm, you have a greater chance of enhancing that rhythm, eventually being able to alter it in the second phase of rider

Close your eyes and have someone lead you.

training, influencing the horse. Identifying the movement of the horse's hind legs will become an essential part of the timing of your seat, weight, and leg cues in the second phase of your training.

The Deep Seat

First, however, you must develop a deep seat, that is, a secure communication between your seat and the horse's back. The more definite and steady this contact is, the more you are going to be able to move with the horse and the less chance there is that you will be jostled out of the saddle during a change of direction or speed. The sensations of muscular coordination that you feel when pumping forward on a swing are very similar to those used when riding with a deep seat. Initially, you might want to identify the

Sometimes it helps to exaggerate in order to find the desirable position. Shoulders rounded forward.

Shoulders pinched back.

Shoulders relaxed.

Collapsed seat.

Hollow seat.

Balanced seat.

physical sensation of the deep seat by pulling yourself down into the saddle with your hands.

Grab onto the pommel of your English saddle or the swells of your western saddle and pull your seat bones down onto the seat of the saddle. You might find it more effective to use one arm in front of the saddle and one on the cantle. With either method, be sure your thighs are relaxed. Tension in your thighs, seat, and back cause you to pinch yourself right out of the saddle, subjecting you to stiff and uncomfortable bouncing. Rather than picturing your legs as a clothespin that holds you onto your horse, think of them as long tentacles that lightly embrace your horse's sides. This will help you achieve a deep seat.

A deep, relaxed seat allows to you follow the movements of the horse in a smooth rolling motion. After you have firmly registered the feeling of the deep seat, gradually take your hands off the saddle and hold them in a position that approximates holding the reins. If you feel you lose the deep, undulating seat, use your hands to pull yourself down onto your seat bones and deep into the saddle once again. Although you will not work exclusively on this exercise, you may need to continue it to some degree for a year or so until you develop a secure, following seat.

Body Awareness: Developing a Balanced Position

Your final goal in the awareness phase of learning to ride is to develop a sense of where all of your body parts are and how they can be used independently. In other words, you need to be able to zero in on your left calf as the horse is moving, sensing where it is located on the horse's side, how much contact it is making with the horse, and so forth. Your instructor will help you identify the location of various parts of your body. After you can focus on any body part, you must learn how to move each part in various ways. Lift the left hand and keep the right hand steady. Bring the right leg back all the way from the hip to the heel. Bring the left leg back just from the knee down. Bring the right shoulder back but not the hand and so on. Again, a good instructor will help you position your natural aids. The effective use of your aids will allow you to maneuver your horse. Just how you influence the horse will be discussed in the chapter on the aids.

One of the first things instructors customarily work on with riders is body alignment. You should pay attention to you posture as part of body awareness. Poor posture during riding can cause discomfort after riding. Proper position is certainly essential for the mastery of riding, but the "head up, heels down, hold it there no matter what" method can create stiffness in a rider. Similarly, fatiguing, repetitious work such as unending circles can serve to make a rider develop worse instead of better posture.

In order to attain proper position, you must use a saddle of the appropriate type and size. The saddle must fit your horse as well as you. If the underlying framework (tree) of the saddle is too narrow, too wide, too long, or too short for your seat, you will have difficulty learning to feel the correct things in order to attain an effective riding seat. The balance of the saddle from front to rear and the location of the deepest part of the seat of the saddle will also affect how successful and comfortable you are when mounted.

If your position is balanced your riding

work should result in minimal stress and wear on your body and should not result in wrenched and twisted muscles. You should not have a painful arm, strained pectorals, or an aching back from riding. Your thighs and abdominals might have that "good soreness" that comes from a vigorous workout, but if your muscles and tendons are stretched, torn, or severely fatigued after riding, something is wrong. Either you are not listening to your instructor, your body, or your horse, or you are working too long.

In order to improve, you must be willing to go the extra mile (sometimes, but not usually, meant literally). Go a bit farther, do a bit more. When you are very tired and ready to quit, ride two more trot-to-canter transitions on a large circle and try to make them the most balanced movements possible, and then finish your session for the day. Although riding should not hurt you, if you approach it diligently it will give you a good muscle tightness in your ribs, abdomen, and thighs. Be prepared to be physically tired and somewhat sore. If you feel strain or pain on your body parts, however, you are likely riding in an improper posture or with tension.

Your riding can become one of the most unified body activities you perform. Some say their body awareness carries over from riding into their other activities. As I sit at my computer, for example, I often become conscious of my posture in terms of the body awareness I have developed through riding. Even though I sit at a desk of optimal height, have a well-designed tilt-and-swivel chair with a sheepskin cover and a lumbar pillow, and use a "good for the back" footstool, I still sometimes find myself sitting imbalanced. I might be bearing down on one seat bone while the other one hovers. My lower

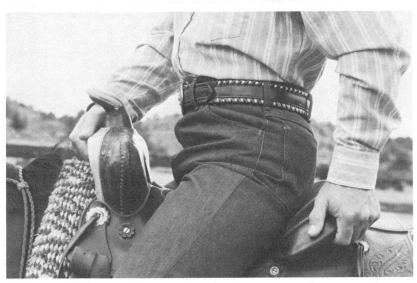

To find a deep seat, pull yourself into the saddle. Western rider.

The beginning rider needs qualified help when developing body awareness so that the right things are practiced.

One of the first things most instructors work on is helping the rider develop an awareness of proper body alignment.

back might be hollow. One side of my rib cage might be collapsed. Most times, however, I find that I am sitting with equal weight on my seat bones and with firm, even pressure of my lower back against the back of the chair; my feet symmetrically positioned either on the floor or the foot stool; my shoulders back, square, and my head on straight! I had not been conscious of the many variations possible just in sitting until I became more aware of my body through riding.

Body awareness can be experienced the other way around, too: awareness during everyday activities can help to improve your riding. One essential for a peak performance, whether it is gliding in and out of your tiny sports car or executing a fluid canter depart, is to find the correct balance and amount of muscle contraction and resistance necessary to get the job done.

Quality of movement is what you should strive to encourage in yourself as well as in your horse. In order to master your horse, you must first master yourself, and that begins with body awareness. Mastery does not indicate brutal dominance — far from it. Instead, think of mastery as control with a subtle fluidity coming from your core and flowing in a circular pathway of energy around you and your horse. Such control is possible only if you learn how to tap the power storehouse you have available within you. The key to your physical performance is located in your mental powers.

Using Mental Imagery to Increase Your Physical Awareness and Decrease Your Tension

As was discussed in the chapter on visualization, you can mentally practice specific skills to help you improve your riding performance when you are not actually riding. The visualization of a skill or a combination of skills can supply you with a mental stimulation that triggers body responses. This allows you to test-run ideas and add to motor skills without actually performing physically. The electrical traces of neuromuscular activity will go through the same patterns as during actual riding, yet the muscular power settings are turned way down so no real muscular contractions take place.

If you are using visualization to help you with a problem and your visual image disappears just when you are about to face that problem, that means you need to work mentally on the individual components to prepare for their proper execution. As you ride in your mind, observe (using your sense of feeling) everything about your body position that could give you clues about what needs work. Are you tense in your back and hips? Do you rise up and off the saddle seat, or do you feel a constant contact between your seat bones and the saddle? Are you pulling back on the reins or holding and balancing yourself with the reins? Or do you feel your hands moving forward in a yielding fashion with the movements of the horse? Imagine you've witnessed a crime and must recall all the important details. In your mental picture of the problem maneuver, what are the important clues?

Mental imagery also allows you to play "what if" scenarios without actually being on the horse. You can run through a variety of happenings that might otherwise catch you unprepared and you can plan and then test your reactions to those situations. You can use mental imagery to help you dispel excess anxiety stimulated by riding, lessons, or competition. You can also use visualization to play "worse-case scenario" about the real risks involved in riding and the fears that cause you physical tension. You can run through your worst fear as many times as you need to until you can look at it calmly and plan what you would do in its event. What do you fear the most? Falling off, getting kicked, being on a runaway horse, or having a horse slip and fall on top of you? Playing such scenes in your mind, where you stay in control and then walk away from the mishap, can do a lot to quell your fears. Learning how to take a fall safely, control a runaway horse, and interpret horse behavior will help you play the scenes with correct details.

MOTOR SKILL LEARNING

Mastering the physical skills involved in riding is a combination of many aspects of learning. Younger people and athletic adults can often just watch a skill and learn it through mimicry. Adults with fewer motor abilities or rusty skills may need to approach motor development another way. First, you must earnestly want to learn the skill. Then you must set a specific, achievable goal and goal plan. Next you need to build the image of the skill in your memory. This is the cognitive phase of motor skill learning, where you can benefit from some knowledge about how to perform the skill.

Once you have a firm idea in your mind of what you are to do, you enter the associative phase of learning where you think less and less about it and focus instead on doing it. Then you practice the skills until they become automatic and you no longer have to think about them. When you are advanced in this phase, your performance looks so easy that an observer has a difficult time figuring out exactly what it is that you are doing.

Learning depends on practice and a knowledge of desired results. Results can be qualitative and subjective, such as an

evaluation of your overall performance by yourself, a judge, or an instructor. Or they can be quantitative or objective, such as achieving a certain speed or getting over a particular fence without knocking it down.

The more precise the knowledge of desired results is, the more rapidly you will progress as a rider. This knowledge of your performance, or feedback, can be provided by yourself, your instructor, a mirror, photos, a videotape, and so on. An instructor who says "good" or "could be better" should be pressed for details. Specific feedback, such as "Your rein is 3 inches too long," is much more valuable.

External feedback serves a motivational function as well. It makes the learner want to practice longer and harder to receive positive feedback in the future. As a rider develops, she designs an internal evaluation mechanism and it becomes a substitute for the instructor's feedback. If a rider is less experienced and has not developed her own means of self-evaluation, then videotaping can be helpful to identify success and failure. For most benefit, the tape should be watched as soon after the ride as possible. After major feedback, a rider may need an interval for absorption before attempting another trial of the skill and getting more feedback. This interval allows the learning to sink in and the rider to make adjustments in her mental pictures. This is learning taking place through visualization.

To achieve the greatest depth of learning, a rider should experience varied practice. Riding alone and accompanied, in diverse conditions, at various locations or settings, in many types of weather and footing, on different horses, and at various times of the day will result in the skill being learned more fully.

Riding is the type of sport that does

Riding in different settings will help you develop the greatest depth of learning.

not lend itself well to strict linear learning. It is impractical to prioritize skills on a list and fully learn one before moving on to the next. If that were the case, beginning riders would work on mounting until they got it perfectly before taking up the reins and asking the horse to walk. Working on various skills of the same level at the same time, with emphasis on the most basic, usually helps a rider progress more quickly and thoroughly.

CHAPTER 7

MINIMIZING THE HAZARDS OF RIDING

Sports challenge the trend of our modern society to remove risk from every aspect of life and make it safe, standardized, and predictable. That is one of the reasons why many people find risky activities such as sky diving, skiing, and riding attractive. The National Center for Disease Control has recently reported that America's 30 million horseback riders suffer a rate of serious injury higher than that of motorcyclists. As with any potentially hazardous activity, from jogging or bicycling along a roadside to stock car driving, every rider must accept her own responsibility for the risks she takes.

Recent legislation in some states (such as Section 13-21-120 of Colorado Revised Statutes, effective July 1, 1990) specifically reminds riders of their personal responsibility for the built-in risks of equine activities. Liability for accident, damage, injury, illness, or death does not fall on a stable, an instructor, or other equine professional, but on the rider herself. It seems odd that legislation was necessary to clarify this for people, yet large lawsuits made liability insurance so costly that riding establishments faced closure unless riders bore liability themselves.

Regulations regarding protective head gear have also been recently formulated. The American Horse Shows Association (AHSA) and the United States Pony Club (USPC) have issued stipulations regarding the type of helmets that are to be used in some of their specific competitions. But the point is not whether or not you participate in competitions that are governed by new regulations. What is most important is that you accept the risks involved in riding and take personal responsibility by preparing and protecting yourself while riding in order to avoid or minimize injury and other harmful effects of riding.

Some of the other hazards, besides injury, encountered while riding are heat exhaustion, sunburn, dehydration, frostbite, and hypothermia. Following recommended horse-handling safety guidelines goes a long way toward preventing injuries. Those rules are discussed in the chapter on lessons. Besides following safe practices, it is advisable that all persons involved in riding take a Red Cross emergency care and cardiopulmonary resuscitation course.

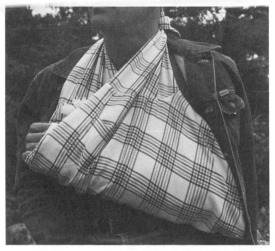

Sprains are another possible riding hazard.

RIDING HAZARDS

Injury

Riding is a physically demanding sport and horses are unpredictable; therefore, injuries occur. Most injuries are minor such as occur in other sports: cuts, scrapes, and sprains.

A common riding injury is the minor inflammation of muscle fibers, the soreness you feel the day after you ride. If you have a general minor tenderness, your symptoms will likely decrease with exercise, which increases circulation and healing. Sometimes you might experience a muscle spasm or a cramp, a muscle contraction that won't quit. Riders most frequently feel these in the calves and hips. Often a cramp is due to muscle fatigue, or it can be due to calcium loss from sweating or exertion. Relax the muscle through careful stretching and massage.

A sprain is the overstretching or tearing of a ligament. A strain is the overstretching of a muscle or tendon. First aid for such injuries usually follows the acronym **RICE**: **R**est, **I**ce, **C**ompression, **E**levation. If you have twisted your ankle when dismounting, your immediate treatment should be to sit down in a comfortable spot with your leg elevated and apply an ice pack to your ankle,

The hazards of riding include abrasions. Know proper first aid.

holding the cold pack firmly in place with moderate, even pressure from an elastic bandage. Elevate your leg so the affected part is higher than your head in order to decrease swelling and pain. Your doctor may recommend you use ice for a cycle of 30 minutes on, 15 minutes off, to be repeated for three hours or more. After a certain period, usually about two days, cold therapy is no longer effective and it may help to switch to heat therapy.

Sun

Ultraviolet and infrared rays from the sun are thought to cause damage to your skin and eyes. Ultraviolet rays reportedly break the skin's collagen molecules and can cause changes that lead to skin cancer. At high altitudes the air is less dense, so the risk of exposure to harmful rays is greater. Squinting from glare causes fatigue in the eyes as well as overall fatigue and can result in errors in judgment.

Heat

If you are suffering from heat stress, you are not going to get the maximum performance from your horse and may incur great bodily damage. Heat stress should be avoided because once someone has suffered it, it seems to recur more easily.

Riders can suffer heat fatigue, heat exhaustion, or heat stroke. Heat fatigue is the mildest form of heat stress and can occur when the body has experienced a 2 to 3 percent weight loss of body fluids. Symptoms are headache, sweat, an increased pulse rate, and an overall tired feeling. Treatment is rest and fluid intake until the water loss has been replenished. The body can absorb 1 quart of water per hour.

Heat exhaustion occurs when the body has a weight loss of up to 8 percent with excessive electrolyte loss as well. The victim will appear pale, sweaty, and weak, may feel cold, clammy, nauseous, and weak in the muscles, and may have a rapid pulse rate. If you feel this way, go to a cool place immediately, elevate your feet and legs, and try to keep your head lower than your trunk. Loosen your clothing and use cold water on your pulse points: wrists, neck, ankles, temples. Drink cool liquids.

A heat stroke victim has lost up to 12 percent of body weight in fluids and requires immediate medical help. The person has a flushed, red appearance with dry, hot skin and no sweat. The body temperature may be as high as 105 to 110 degrees Fahrenheit. Other symptoms include a complete loss of muscle tone, pinpoint pupils, and deep rapid breathing. The victim may enter a coma. Until he or she can be seen by a doctor, the body should be cooled by ice. People with a

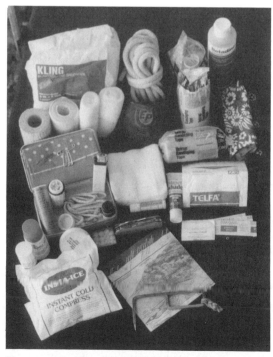

The contents of a first aid kit for a saddle bag.

high fat content in their bodies are more prone to heat stroke as the fat insulates and keeps heat in. Also those persons who work and live in air-conditioned environments and then exercise vigorously out in hot temperatures are more likely to suffer heat stress.

Developing heat tolerance. Any athlete must build a tolerance to heat by adaptive acclimatization and a program designed to increase the body's ability to dissipate the heat of exertion. As the body becomes more efficient at handling heat the following changes take place:

- the body begins to sweat at a lower body temperature

- sweat glands become more proficient, producing more sweat

- the number and size of surface capillaries increase to assist in evaporative cooling

- the amount of salt in the sweat decreases by as much as ten times

To become more accustomed to heat, gradually increase the amount of time you ride in the heat and increase the amount of energy you expend in hotter temperatures. Be sure to keep your water intake high, as dehydration leads to heat stress.

Dehydration. Body fluids are lost through sweat, urine, and respiration. Very strenuous riding on a hot day may deplete your body of 1 gallon of fluid per hour, which could set the stage for dehydration. A 2 percent weight loss for a 150-pound person would be 3 pounds or 1½ quarts. Besides losing fluids, your body also loses important electrolytes such as potassium, sodium, calcium, and magnesium. Lack of some of these electrolytes can cause fatigue, muscle cramps, and light-headedness.

Drink plenty of water to avoid dehydration.

The color of your urine will tell you if you need to drink more fluids. If it is darker than the color of apple juice, it is fairly concentrated and you need to drink more water. If it is clear, you are probably taking in enough water to flush the soluble wastes from your body.

Electrolyte supplementation? Too much salt can cause your feet and ankles to swell and can increase your blood pressure. Prescribing salt tablets to athletes before, during, or after exertion does not hold water. Well, in fact it does "hold water"—and that is why it is undesirable. Salt tablets will cause an increase of water absorption in the gut, making less available in the bloodstream where it is needed.

Humidity

Humidity slows or stops evaporation and so compounds heat stress by making your body's cooling system less efficient. Air at 60 percent humidity means that the air is holding 60 percent of the water it can hold. The combined effect of temperature and humidity can have a great effect on the performance of both the equine and human athlete. A combined temperature and humidity value over 150 (especially if the humidity contributes one half of the sum) may make it difficult for a rider to self-cool through perspiration alone. External cooling such as an ice-water bandana around the forehead or neck is helpful. Working in a heat/humidity combination over 180 can be dangerous.

Cold

The effect of the cold temperatures of the fall, winter, and spring can be compounded by wind chill, humidity, precipitation, and cloud cover. The wind chill is the speed of the wind in relation to the temperature. So a 30-mile-per-hour wind at 10 degrees Fahrenheit feels like a minus-33-degree temperature with no wind.

During cold temperatures the body's surface capillaries constrict so that heat cannot escape through the skin. This is what makes the hands cold and stiff. If carried to extremes, this can result in frostbite, most often of fingers, toes, ears, and nose. First aid for frostbite is to get immediately to cover and warmth. Take in warm fluids. Although you should not rub frostbitten skin, you can immerse it in warm, not hot, water.

Hypothermia, or extreme body cooling, occasionally occurs when a rider is injured or stranded in a storm. Hypothermia is initially caused by low environmental temperatures, precipitation, and wind, and can be compounded by hunger, fatigue, or exhaustion. The body continues to burn fuel to generate internal heat and when the fuel reserves get low, an emergency situation sets in. Some of the symptoms of hypothermia are shivering, rapid muscle contractions, clumsiness, mental confusion, sleepiness, and then an indifference to what is happening. Hypothermia is a medical emergency requiring a doctor's attention. The victim should be taken to a shelter, given dry clothes and a source of external heat (fire, hot water bottles, another person's body heat), and given warm liquids until a doctor can take over.

Altitude

Riding in high altitudes has the most effect on endurance activities because the reduced air pressure at high altitudes decreases the amount of oxygen available to the blood cells for energy. It may take you (and your horse) one to two days to acclimatize to altitudes above 7,000 feet if you are accustomed to working at sea level. Riders improperly acclimatized may suffer altitude sickness, feeling lethargic and drowsy with headaches and have difficulty sleeping. It has been found that to minimize these symptoms, a person should follow a diet high in complex carbohydrates and low in fat and not consume alcoholic beverages.

Higher altitudes do offer some benefits. The reduced air density of "thin" mountain air decreases the resistance in the airways to air flow. Also, the air is usually cooler and drier, which can help prolong work and maximize body cooling. In any environment with a low humidity, fluid intake must be high to prevent dehydration.

THE RIDER'S DIET

The rider's diet must be balanced for maximum performance. A knowledge of nutrients will assist you in making wise diet strategies to reach your riding goals. Carbohydrates (sugars and starches) are the main source of food energy. The simple sugar glucose can go directly from the intestine to the bloodstream. Compound sugars such as fructose and galactose must go to the liver to be changed into glucose before being used or stored as glycogen. Therefore, if you need an immediate energy source, you would do best to choose something containing glucose. If you are eating to fuel up for a moderate to long ride, a complex carbohydrate would be a better bet.

Fats are also a source of energy; in fact, they are the main source of stored energy in the body. Fat provides two and a half times the energy (calories) of an equal weight of carbohydrates. You don't have to eat fat to store it because nearly everything you eat can be converted to and stored as fat. Body fat performs some important functions: it cushions and protects organs; it insulates against the cold; and it stores fat-soluble vitamins A, D, E, and K. Your diet should therefore contain between 10 and 20 percent fat, most of it to be used for energy fuel.

Protein, essential for the formation, growth, and maintenance of cells, can be changed into carbohydrates or fats to be used as energy or storage, depending on the body's needs. Protein is broken down into amino acids, the building blocks of the body. Twenty-two different amino acids are necessary for proper functioning of the human body. You can manufacture fourteen of these amino acids from the food you eat in a balanced diet. The other eight essential amino acids, however, must be obtained daily, directly from foods. Those foods that are termed complete proteins—meat, eggs, and milk—contain all of the essential amino acids. Some foods contain complementary proteins and when combined, they result in complete protein. Examples are wheat plus beans; rice plus beans; nuts plus grains; and potatoes plus milk products.

Water, which makes up 60 percent of your body, is essential for all bodily functions. You should drink four to eight glasses of water a day depending on your climate and level of exercise. Beware of substituting other liquids (such as coffee, pop, juice, or milk) for the majority of your fluid requirements as you may be getting undesired or unneeded substances (caffeine, sugar, fat, calories) along with the fluid.

Vitamins are organic substances necessary to many bodily functions. The fat-soluble vitamins, A, D, E, and K, are stored, so rarely are deficient. Unwarranted over-supplementation can cause a toxic overdose. The water-soluble vitamins, such as the B complex and vitamin C, are required daily as the excess is flushed out in the urine. Adequate vitamins are usually provided in the well-balanced diet.

Minerals are inorganic micro-nutrients important for maintaining the electrolyte balance crucial to the body fluid balance and the cooling system. Sodium is the largest mineral constituent of the blood. Although the body needs sodium chloride to retain water, the average U.S. diet contains fifty times more salt than needed, so you should never need to take salt tablets unless ordered to by a doctor. Potassium acts to widen blood vessels to enable the body to better cool itself by evaporative cooling. When your potassium levels are low, you can become tired and run-down. Potassium is common in fruits and vegetables and is a major constituent of "lite" salt. Calcium is lost in sweat and a low calcium level can result in muscle cramps.

A Rider's Diet Strategies

When you are preparing for active work, do not tax your intestines with the wrong foods. Before your daily ride, lessons, or competitions, follow these healthy rules.

1. Avoid caffeine drinks. Too much caffeine excites the nervous system and later produces a mild physical and mental depression. Caffeine is also a diuretic, robbing you of precious fluids and electrolytes.

2. Avoid sugar. In a typical can of soda there are five teaspoons of sugar. Sugar is a common American addiction, thought to be responsible for tooth decay, kidney disease, heart disease, high blood pressure, the onset of diabetes, and behavior disorders such as the "sugar blues," the depression that follows a sugar let-down. If you eat a candy bar and drink a can of pop one-half hour before you start the cross-country phase of your competition, you may not only find you have lost all of your gas by the fourth or fifth fence, but that you are too light-headed and weak to continue.

3. Eat complex carbohydrates in moderate amounts several hours before performance time and possibly a small amount of simple carbohydrates one hour before. Examples of complex carbohydrates are pasta, grains, breads, and potatoes. Examples of simple carbo-

hydrates are fruits and fruit juices. Right after you eat, blood is shunted to your gut to aid digestion, so your skeletal muscles contain less blood than usual and therefore cannot perform at their peak. But if you schedule your meal several hours or so before performance, your blood has time to return to your skeletal muscles so they can be freshly fueled.

4. Don't eat vegetables or fats as part of your pre-performance meal. Salads, green vegetables, and fats take longer to digest than do carbohydrates and protein. Although vegetables and fats are an essential part of your daily diet, eating them prior to a performance may cause digestive discomfort.

5. Don't drink huge amounts of liquid prior to performance. Only about one quart of liquid per hour can leave your stomach; the rest will be carried around with you as you perform. Many frequent sips before, during, and after riding are best to prevent bloating.

6. Don't consume alcohol before you ride or while you ride and use only moderate amounts of alcohol in your routine diet. Alcohol is a poor source of energy but a major source of empty calories. At first alcohol may seem like a stimulant, but in fact it is a central nervous system depressant. It dehydrates the body by widening the blood vessels, increasing evaporation and increasing heat loss. The warm feeling that alcohol gives is a sure sign that heat has moved to the surface of the body and is escaping. Alcohol impairs reaction time and judgment and is dangerous around horses. In addition to the physiological reasons that an athlete should not use alcohol, it has a high incidence of being psychologically addicting as well.

SPECIAL CONSIDERATIONS FOR THE OLDER RIDER

There is no reason why you can't continue your riding as long as you can walk. Most changes attributed to age are really due to a lack of conditioning caused by low physical activity levels. If you are inactive, riding, like any athletic pursuit, may put a strain on your body. The pattern usually goes something like this: increased age, decreased activity, decreased caloric requirements, equal or greater caloric intake, increased weight. A weight gain can affect many body functions in a negative fashion, yet the exercise characteristic of riding and its associated horse-care tasks can help you keep an optimal weight as well as provide other benefits: lower blood pressure, regulate blood sugar levels, increase efficiency of the heart and circulatory system, decrease blood cholesterol, and allow higher-quality rest.

Living a "life of convenience" can exaggerate the changes attributed to age. Sedentary individuals may experience musculoskeletal changes due to the long hours of sitting. With disuse, muscles lose tone and size and bones lose density. Some of the possible conditions that riders over forty should pay special attention to are osteoporosis, arthritis, lower back problems, and tendinitis. Arthritis sufferers often benefit from specific stretching and strengthening exercises that help take the stress off bones or affected joints. Tendinitis, the inflammation of a tendon, is often very painful, yet some experts advocate strengthening the muscle to alleviate the pain. Riding can provide the right type of exercise to minimize the effects of physiological aging.

Lower back problems are sometimes due to lack of exercise, too much sitting, muscular imbalances, and poor posture. Balanced riding can improve all of these things. Pursuing riding in an improper manner, however, can and probably will intensify musculoskeletal problems. The older rider needs to pay close attention to warm-up, cool-down, correct riding, and exercise between rides.

CHAPTER 8

CLOTHING FOR RIDING

Choosing the right clothing for riding will add to your safety, comfort, and effectiveness.

SAFETY EQUIPMENT

Protective Helmets

Beginner riders, young riders, and all riders participating in events involving speed, jumping, or cross-country work should seriously consider wearing a protective helmet. The helmet should be well made, properly fitted, and secured with a chin strap or harness. The recent action by the AHSA and USPC requiring helmets has resulted in the design and testing of new riding safety helmets that are supposed to reduce the impact to the rider's head from the ground, a horse's hoof, fences, or other objects. To create a performance standard, the USPC consulted with the American Society for

Testing Materials (ASTM), which sets technical standards for a wide variety of products; the standard for protective riding helmets is ASTM F1163-88.

Manufacturers of helmets can voluntarily submit their helmets to be tested by an independent firm, the United States Testing Company, Inc. Helmets that are certified by USTC to meet ASTM standards are labeled with an approval from the Safety Equipment Institute (SEI), a private, nonprofit organization. Some of the changes and improvements designed for the new helmets include:

- *Improved impact absorption.* The helmet should be able to absorb a heavy impact without transmitting more than 300 g's (g = the force of gravity) to the rider's skull and brain. 300 g's is the generally accepted level of acceleration beyond which concussion (not necessarily fracture) will occur. This is three times the previous standard of protection for riding helmets. The increased protection may be provided by an expanded styrofoam liner, replacing the previously used foam rubber. Foam rubber is thought to collapse initially and then bunch, transmitting energy to and traumatizing the brain. Styrofoam is said to absorb energy uniformly over its entire depth and transmits less energy to the brain. Although it feels hard, it should cushion three times better than foam rubber. Styrofoam is not resilient, however, and will not regain its original shape, so after receiving a sharp blow, the liner will need to be replaced.

Use a well-made helmet to protect against head injury. Here it is outfitted with ear flaps, neck gaiter, and glacier glasses for winter riding.

- *Increased retention strength.* The permanently attached chin strap and harness have been redesigned to meet a severe downward load without failure and should not come off during a fall.

- *Similar area of coverage.* The helmet passes above the ears and curves down at the back of the neck.

- *Increased external size.* The overall diameter is ½ inch larger than previous models to accommodate the liner material needed to meet the impact absorption standards.

Other Protective Equipment

- *Gloves.* Using gloves while working with horses provides a variety of benefits including keeping the hands clean and warm and protecting the skin from abrasions such as rope burns.

- *Pants.* Riding pants should allow freedom of movement without being baggy. Loose wrinkles can cause painful raw spots on your skin. Many types of stretch jeans and breeches are specially suited to riding. Choose those with minimal bulk at seams to decrease the chance of rub marks. Some riders like to use pants with leather knee patches or full leather seats as well as full leather chaps. Although these can serve to stabilize a rider's position, the increased friction may make it difficult for a developing rider to adjust his or her position.

- *Boots.* Boots for working around horses on the ground should have good traction and be constructed of material that will protect the foot if it is stepped on. Sandals, thongs, or light canvas shoes should never be worn around horses. Some leather athletic shoes (running, jogging, or hiking) are made with enough substance to provide adequate protection for ground work. For riding, however, always choose sturdy shoes or boots with heels to help prevent the foot from slipping though the stirrup. Leather soles can be slippery on dry grass and deteriorate rapidly when subjected to mud and manure. Be wary of using footwear with heavily lugged soles as these may cause your foot to get hung up in the stirrup, especially when using stirrup irons. To prevent the stirrup leathers from pinching your leg while riding English, choose knee-high boots or use suede gaiters with shorter boots.

- *Hair accessories.* If you have long hair, secure it back, away from your face and line of vision. As well as keeping it out of your eyes, this will minimize damage from sun, wind, and dryness. For horse shows, hair should be neatly arranged so as to add to the impression of care and control. Experiment with barrettes, hair bands, sweat bands, hair nets, hats, and snoods to find what works the best for you.

- *Sports underwear.* To protect breast tissues from the constant jarring caused by the horse's movement, especially trotting, female riders should consider using use a supportive "jog bra" specially designed to hold the breasts close to the rib cage. Male riders should consider using an athletic supporter.

- *Rain gear.* During wet weather protect yourself and your saddle with a long, water-proof coat.

Clothing for Summer Riding

As you choose your summer clothing, keep some general rules in mind. Light-colored clothing reflects the sun's rays, and dark clothing absorbs them. Loose, open weaves allow for continual gradual cooling. The type of fiber and the weave have a lot more to do with how cool you feel than whether or not you are wearing long sleeves. You would probably feel much hotter in a tightly woven polyester short-sleeved polo shirt than in a loose, long-sleeved, mesh cotton shirt. Cotton is a favorite summer fiber for active riders. Breeches and jeans made of cotton or Cool-Max fabrics are much more comfortable than Lycra/Spandex pants or polyester jeans.

Although exposed skin allows for more evaporative cooling than covered skin, it also absorbs more ultraviolet rays. Although a bit of sun can bring color to your cheeks, too much can be destructive to the skin. Get in the habit of wearing a brimmed hat for sun protection. Use high-quality sunscreen regularly on your face, hands, and lips. Take extra good care of

Avoid summer's damaging rays with a broad-brimmed hat, sunglasses, a protective neck scarf, and a snood.

your eyes by making a pair of high-quality sun glasses an integral part of your riding outfit. Buy the ones that provide at least 98 percent protection from ultraviolet rays. Tinted lenses with less protection from UV rays may be worse than no sunglasses at all because the dark lenses encourage the pupils to dilate, making them more susceptible to harmful glare.

Clothing for Winter Riding

One of the primary problems facing winter riders is how to keep the extremities warm. A few common-sense rules, however, coupled with today's vast selection of high-tech, innovative clothing, will guarantee your comfort and may turn winter into one of your favorite riding seasons.

Although clothing preference is largely influenced by individual taste and the performance of the clothing is greatly affected by personal body chemistry, certain general principles apply to all cold-weather garments.

Choose fabrics carefully. The accompanying materials charts list the natural and synthetic fibers most commonly found in riding clothing along with their fiber characteristics and care. While this fiber information will provide a base, ultimately you will have to do your own testing. You'll probably find that certain garments make you feel so toasty and comfortable that you'll want to wear them for more than riding. Other items may not allow you to cool properly, may cause itching, or may be generally uncomfortable.

Design your winter wear so that perspiration has a place to go. Choose garments with necklines that can be adjusted with zippers or buttons to vent

Winter clothing.

body heat. Fibers and clothing materials either transmit, repel, or absorb moisture. Select fabrics for your undergarments that transmit or "wick" moisture away from your body. Materials such as cotton that merely absorb moisture become soggy and heavy. Even worse, they hold the moisture next to your body, making you colder. Materials that repel moisture do not let moisture in or out so perspiration remains on your skin.

Wicking is a process whereby perspiration is transported from the body's skin to the underclothing and then to other layers of clothing or out to the air. This can happen in several ways. Some fabrics with close-together fibers, such as silk, allow sweat to "climb" the fibers by capillary action to the outerwear. Open-weave, non-absorbing fabrics also remove sweat from the body by allowing body vapors to evaporate outward through the spaces in the weave. A third type of wicking is "spreading action," as in the treated polyester, Capilene, whereby water molecules are attracted to certain types of dry fibers more than they are to other water molecules. The moisture migrates through the clothing, always seeking the driest fibers, moving to the outer layers and eventually dissipating.

Dress in layers. Layering traps air, which helps to keep cold out and body heat in. Proper layering techniques prevent overheating, soggy clothes, and subsequent chill. If you layer, you can peel off your clothing to accommodate the various activity levels during your riding session. In simplest terms, the necessary layers for winter riding consist of an under layer to handle perspiration, a middle layer to provide insulation, and an outer layer to cut the wind and protect from precipitation, if necessary.

Dress according to your activity level. Rate your riding as passive, active, or very active. Quiet trail riding at the walk is passive, so extremities (fingers and toes) may get cold from decreased circulation. Cross-country galloping is very active for the rider and rarely results in coldness during reasonable environmental conditions. If you are cold, you can increase your activity level. Move from a jog into a posting trot, or get off and lead your horse to restore the blood flow in your legs and arms. If you are too warm, temporarily decrease your activity level and gradually vent body heat. A too-rapid evaporation of body moisture and heat can lead to chilling, so adjust your clothing in small stages rather than all at once.

Carry compact, lightweight accessories. It doesn't matter if you are headed out to your arena or down the trail, always take along added protection such as neck gaiters, scarfs, and extra gloves or socks. These things take little space in your pockets or saddle bags but can make a real difference in your comfort. If you are going to be more than a few minutes away from shelter, tie a storm coat to your saddle.

Avoid tight clothing. Pressure on surface capillaries from tight hats, boots, gloves, or clothing will slow circulation and decrease body warmth. Loose clothing traps fluffy layers of air, which then act as insulation.

PROBLEM AREAS

Here is some specific advice for problem areas common to riders.

Feet. The important layering principle starts at ground level. Use a thin sock liner (silk, Thermax, or Capilene) in conjunction with a medium-weight wool outer-sock. This combination provides wicking, insulation, and cushion. The thickness of your outer sock will be determined by your activity level, the insulating quality of your boots, and boot fit. Be sure to

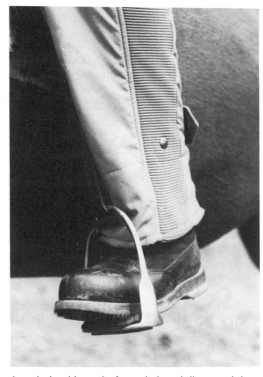

Insulated boots for winter riding; rubber stirrup treads.

choose boots that fit well but are not the least bit tight. Even a slight pressure at the toe or instep will hamper circulation and result in cold feet, no matter how expensive your socks are. Good-quality insulated boots have an inner layer of perspiration-wicking material. Often the insole is removable to allow drying. The middle layer is made of selectively permeable material to act as a mediator between your foot's perspiration and the snow and slush. The outer layer should be a durable yet porous covering, such as leather or Cordura Nylon. Canvas or nylon snow-packs (usually with removable felt liners) are warm and roomy but may be too wide to use safely in a stirrup and some have poor traction on frozen ground.

Just as a Thermos can keep both lemonade cold and coffee hot, insulated clothing functions similarly. If you put a cold clammy foot into a cold insulated boot, that boot will tend to maintain the cold temperature of the foot and boot materials. Your circulatory system, which provides internal heat to your body via your blood, will have to work harder to warm up both your cold foot and the cold boot. So start with warm, dry boots, warm feet, and be sure your circulation is not impaired by either tight boots or socks.

If you are riding English, use stirrup treads with your irons. The insulation of the rubber treads prevents the cold metal irons from drawing precious heat from your feet. If you are riding western during very cold weather, you may wish to consider using lined *tapaderos*. These cozy toe covers can make a big difference in foot warmth provided that the boots you are wearing are dry and snow-free and your feet are relatively warm to begin with.

Hands. If your winter riding does not require intricate rein aids, there are many fluffy, lightweight polyester-filled gloves that you can use. If you need to ride with more precise contact on the reins, using a pair of silk, Capilene, or Thermax glove liners under a pair of slightly over-sized riding gloves works well — either leather or nylon with rubber-pimpled palms. For temperatures around 30 degrees Fahrenheit, cross-country ski gloves, often a blend of warm fabrics and leather, provide grip and warmth.

Head. Don't be casual or negligent in your choice of headgear as you can lose a great deal of precious body heat through your head. When the weather is below freezing, and especially if there is wind, choose an insulated hat or helmet with a tightly woven outer cover to deter heat loss from convection by the wind. Hats or helmets with optional ear and forehead flaps prove their worth when the weather changes for the worse mid-ride.

Heavy sheepskin hat, neck gaiter, and jacket with adjustable neck.

During storms a western hat's brim may protect you from falling snow, but meanwhile your ears are freezing. You can use your western hat with either a hood/neck gaiter, a ski balaclava, or specialized western hat ear flaps.

Just because it is winter, don't forget about the sun. In fact, you may be receiving an increased dose of harmful ultraviolet rays from the glare of the sun on snow. Brimmed hats provide some protection from above but you still need to use sunscreen during the winter. Continue wearing high-quality sunglasses throughout the winter.

Neck. Articles of clothing around your neck should be adjustable or removable so you can retain or vent body heat as desired. Jackets with collars that zip up to your chin allow for a great range of adjustment. Zippered turtlenecks will allow you to release body heat during very active riding. Neck gaiters and silk scarfs can be worn over or under your jacket and can be removed and stuffed into a pocket when no longer needed.

Legs. For very cold weather and passive to moderately active riding you may wish to use insulated pants that are specifically designed for riding. If you want to wear jeans or breeches, choose a roomy pair and wear them over drawers made of silk or polyester. To keep the bottoms of your pants dry and your boots from filling up with snow (especially if you will be leading your horse), use gaiters over your pants and boots.

Upper body. A rider's trunk rarely gets cold because it benefits from the heat of body metabolism (circulation, respiration, digestion). That is why you must pay particular attention to layering the upper body. First select an underlayer (silk, Thermax, or Capilene) to suit your activity level. Then cover it with one or more insulating layers (Synchilla, wool, down, Thinsulate). Top this with a wind-resistant yet breathable layer (nylon, densely woven cotton). Some jackets perform the functions of both the middle and outer layers. During temperatures in the 30s and 40s, when a jacket may not be necessary, a good combination is a silk turtleneck, wool sweater, and tightly woven riding vest. A nylon windbreaker, preferably the type that can be stuffed into its own pocket and snapped to the saddle, repeatedly proves its worth.

Outer garments that unzip or unsnap from both the top and the bottom are most suitable for riding. Be sure your riding jacket allows freedom of movement in the shoulder and arm areas. Longer coats should have flaps or gussets that conform to your sitting position and the saddle's cantle.

To add to your comfort in the saddle during winter riding, make or purchase a sheepskin saddle seat cover. The air trapped between the fibers of the sheepskin warms up much more quickly than does the stiff cold leather of an uncovered saddle seat.

Fibers and Materials

Key:
a. Fiber or material
b. Definition
c. Characteristics
d. Care

1. a. Cambrelle
    b. synthetic fiber used as lining in boots
    c. absorbs foot perspiration, dries quickly, resists odor and mildew
    d. let dry at room temperature

2. a. Capilene
    b. Patagonia's registered polyester fiber
    c. outer skin of each fiber treated to attract water (transmits sweat and allows wash water to remove salt residues) while inner core repels it; won't stain or absorb oils or odors
    d. machine wash and dry

3. a. chlorofibre vinyon
    b. a synthetic fiber used in clothing
    c. very soft, very comfortable, very high heat retention, good wick-ability; if not blended with other fibers, can stretch out of shape
    d. machine or hand wash, drip dry

4. a. Cordura
    b. a heavy, textured nylon often used as an outer layer
    c. lightweight, durable, abrasion-resistant, rot and mildew proof
    d. virtually care-free; brush or wipe

5. a. cotton
    b. fibers from bolls of cotton plant; fibers ¾" to 2½" long; 90% cellulose.
    c. absorbs perspiration and does not retain insulating properties when wet; either retains moisture next to skin or evaporates quickly, causing body to cool
    d. machine wash and dry but can shrink

6. a. down
    b. soft small feathers from the skin of a bird
    c. light, resilient, and good insulation depending on quality; may be expensive
    d. may need to be dry cleaned

7. a. Gore-Tex
    b. Teflon membrane with 9 billion pores per square inch, each pore 2,000 times smaller than a drop of water but 900 times larger than a molecule of body vapor
    c. blocks water penetration but allows for escape of perspiration; Is windproof, waterproof, and breathable
    d. machine wash and dry

Fibers and Materials (cont.)

8.   a.   leather/skins
     b.   preserved animal hides without hair or skins with hair
     c.   protects from wind but permeable by air and water; resists abrasion and tearing; long lasting; can be heavy
     d.   depending on article, dry clean or wash with mild soap and oil occasionally

9.   a.   nylon
     b.   synthetic plastic material
     c.   strong, abrasion- and tear-resistant; long-wearing; will not shrink, mildew or rot; generally not permeable by air
     d.   virtually care free; brush or wipe

10. a.   polyester (See Capilene, Synchilla, Thermax and Thinsulate)
     b.   long-chain synthetic fiber
     c.   low moisture absorption; high elastic recovery; can accumulate static
     d.   machine wash and dry on low or dry clean

11. a.   polypropylene
     b.   synthetic fiber used in under-garments
     c.   fibers do not absorb water; surface of fibers pitted, therefore difficult to remove body oils and odors; can melt
     d.   machine wash, drip dry — no heat

12. a.   silk
     b.   filaments from cocoons of silkworm moth
     c.   single continuous strand can be 2,000 to 3,000 feet long; yarn is comprised of several filaments; absorbs perspiration up to 33% of its weight without feeling damp; lustrous, smooth surface that doesn't retain soil or odor
     d.   hand or gentle machine cold wash, dry flat; or dry clean

13. a.   Synchilla
     b.   Patagonia's registered name for lightweight polyester pile fabric. Pile is a cloth whose surface has a great number of threads which stand erect from the foundation structure of the cloth, therefore is very fluffy
     c.   absorbs very little water, dries out very quickly; has a very high warmth-to-weight ratio
     d.   machine wash and dry on low

14. a.   Thermax
     b.   100% polyester from DuPont
     c.   hollow fibers; wicks moisture away from body but doesn't absorb it
     d.   machine wash and dry

15. a. Thermolactyl
    b. Damart's blend of chlorofibre vinyon and acrylic yarn
    c. absorbs perspiration; transmits moisture to outer layers to evaporate; extremely comfortable; good shape retention; resists odors
    d. machine wash, drip dry

16. a. Thinsulate
    b. thermal insulation: Type B is comprised of 100% olefin, a paraffin-based synthetic; Type C, of 65% olefin, 35% polyester
    c. micro-fibers, some ten times smaller than polyester fibers, create large surface area to trap air for insulation; absorb less than one percent of water by weight
    d. machine wash and dry

17. a. Vibram
    b. boot sole made of rubber compound
    c. originally designed as a lug sole for climbing boots, now Vibram soles come with a variety of surfaces; provides cushion, protection, grip, protection from snow and water
    d. wash or brush

18. a. wool
    b. fabric knit or woven from yarn spun from the fleece of sheep
    c. a relatively coarse fiber 1 to 6 inches long; lightweight; absorbs as much as 18% of its weight; release of absorbed moisture is gradual so wool is slow to feel damp and does not chill wearer by too-rapid drying; retains some of its insulating properties when wet
    d. hand wash or mild-action cool machine wash; drip dry

SECTION THREE

THE RIDING

CHAPTER 9

RIDING CONCEPTS

As you work with a horse, you will become more aware of your own inherent and conditioned reflexes. To make things more interesting and perhaps more complicated, you will also need to become very familiar with your horse's reflexes. As your reflexes are being identified and honed for riding, your horse's reflexes are also going through a similar process of identification and modification.

In a very general sense, your goal in influencing a horse is to have him do what you want him to do. Oftentimes this is the very opposite of what the horse would instinctively choose to do. In fact, a point of humor among professionals is that we discover what the horse does *not* want to do and then we train him to do just that! There is a lot of truth to the jest, except in the case of a naturally talented horse that is already mentally and physically designed to do what we wish. However, horse training is an enormous separate subject; this book is devoted to the training of the rider. In my discussion, therefore, I will assume that the horse you are learning

on is obedient and well trained — that he already does *what* you ask.

When horsemen define *how* a horse does what we want him to do it is necessary to talk about the quality of the horse's actions. To give you the best chance for a comfortable ride, a horse should move with forward energy, rhythm, balance, and smoothness.

Forward energy. Active, forward energy is essential for all gaits and transitions. It is difficult to ride well if you must continually prod a horse along. A horse that is eager to step forward with his hind legs will be easier to shape into both upward and downward transitions. Such a horse will feel like he is stepping up into your hands. A horse without sufficient forward energy may learn to back away from the contact with the bridle and move with short, tense steps. Or he may move sloppily, fall apart in the middle, and make you feel as if you are riding two separate horses — one in front of you and one behind. The energy the horse generates with his hind end provides you with the means to tie all of your aids together and increases your chance for a unified movement with the horse.

Rhythm. A regular rhythm will offer you a very predictable set of movements. When you can depend on a horse to move his legs in a precise and very even two-beat, three-beat, or four-beat time, you will find it much easier to stay with his movement, find your seat, and stabilize your hands. Rhythm is inherent to some degree but can be further developed in both the horse and rider through practice.

Balance. The definition of balance refers primarily to the working relationship between the front and rear portions of the horse. The majority of a horse's weight is in his forehand. Left to his own devices he would probably travel more heavily on his forehand because of the simple fact that it is easier. It requires less work on his part to leave the weight forward where it naturally settles, rather than lifting it up with each stride, rebalancing it to the rear, and carrying it with the hindquarters. When you ride a horse heavy on the forehand, you will find that such a manner of travel is not comfortable. You will probably feel like you are sliding downhill or your arms and upper body are being pulled forward, which in turn lifts your seat out of the saddle.

Your goal is thus to convince the horse to shift his balance rearward and carry more of his weight with his hind legs. This will give you a more comfortable spot to sit — a still spot in the center of the motion of the horse's body, something like the quiet at the center of a teeter-totter. To provide riders with this quiet spot to sit, most horses must be convinced that they can and should work harder and that they can carry more of their weight and ours with their hindquarters. This takes training and conditioning so that the horse can

develop the necessary muscles to be able to carry himself in this balance. When we have achieved this with a horse, we have positively altered the horse's anterior/posterior balance.

There is also the matter of left-to-right balance to consider. Just as humans usually have a side preference, so do horses. That is why some horses travel straight and fluid in one direction and crooked or stiff in the other. A crooked horse tends to throw a rider into a crooked position, such as a collapsed hip or twisted shoulders. Ideally, the horse's body should travel in a relatively straight line and should remain even from side to side when turning both left and right (see more on lateral balance at chapter end). A straight and even horse will contribute to your comfort and correctness when you ride.

Smoothness. Besides energy, rhythm, and balance, one other goal for comfortable riding is smoothness. It is difficult for any rider to sit well on a horse that falls or leaps into a canter or drops its weight onto the inside shoulder on a turn. Smoothness starts with a well-trained, well-conformed horse that has a relaxed mind and a relatively fit body. Avid riders aim to make all movements and changes in movements (transitions) so subtle and smooth that they are barely perceptible to the rider's body or to an observer's sight.

RIDING AND TRAINING TERMINOLOGY

You may have noticed that horse people speak in a specialized tongue. Horse jargon and, in this case, riding and training jargon, is composed of words that don't provide a clear, accurate picture by themselves. They require a detailed definition or explanation. (See Glossary.) Sometimes a single word can have several meanings and a host of applications and interpretations, depending on the style of riding being discussed. Following are some important riding concepts with their classic, broadly applicable meanings.

Freedom of movement, free spirit. When applied to horses, the word "free" confuses some people. Freedom indicates a lack of restriction and usually brings to mind a horse galloping through a field. Yet free is not necessarily synonymous with wild. A well-trained and disciplined horse can and should move freely. This means using aids that are effective without being forceful. The freely moving horse shows expression in his face, body carriage, and the way he lifts his legs and places them down. To be able to ride a horse freely yet totally under control is the ultimate goal of riding, achieved by the advanced rider. To allow a horse the freedom to express, the aids must be applied in the appropriate position, at the optimal time, and with the optimal intensity. It takes years for a rider to develop the ability to influence a horse without restricting or blocking the energy flowing around the horse's body.

Harmony. This describes a good working relationship between horse and rider, one with a smooth flow of energy and an open line of communication. To achieve harmony, you may need to assess your own state of mind and admit when changes in attitude might have to take place in order to have a productive ride. Harmony is evident in the carriage of both the horse and rider. Poise, confidence, and pride in work are characteristics of horse and rider harmony. Harmony is a state of relaxed energy. If a rider is tense,

Freedom of movement means riding a horse with aids that are effective without being forceful. This horse is apprehensive about approaching unfamiliar objects, and the rider's aids are light and effective.

she will block the horse in some way and the horse will undoubtedly become tense as well.

Connection. The connection is the cooperative relationship and balance between the rider's driving aids and her restraining aids. The rider uses her seat, legs, upper body, and weight to encourage a horse to move forward and she uses her seat, lower legs, and hands to shape the energy she has created with the driving aids. That solid, uninterrupted feeling of energy flow through the rider's and horse's body is the connection. It is as if the rider is plugged into the horse. The well-connected horse drives powerfully yet fluidly from the hindquarters, sending the energy up to a "well-hinged" forehand — one where the neck is suspended from and carried by the upper neck muscles, which originate at the withers.

Contact. Contact refers to the pressure felt by both horse and rider through seat, legs, and reins. In its most common usage, contact refers to the relationship between the horse's mouth and the rider's hands, via the reins. Contact should be even and

The "connection" is the flow of energy around the horse and rider.

steady. Wavering contact, caused by even a slight movement of the rider's hands, creates a bumping in the horse's mouth.

The degree of contact will vary according to the riding style and the level of the horse's training. Commonly used phrases describing amount of contact include "a light feel," "on the bit," "in the bridle," and "up in the bridle." Riders using a slack rein, as is characteristic of western riding, aim for a very light contact, using just the weight of the reins on the sides of the horse's neck and small movements of the reins to the bit. The reason western horses can be controlled on a slack rein is because they have been trained to respond to the rider's weight, seat, and legs as primary cues.

Horses ridden with a shorter rein, often associated with English riding, will feel a greater degree of constant pressure on their mouths from the action of the bit. Theoretically, this allows the rider to make more subtle adjustments in the reins to cue the horse. Riding with too strong a contact on the reins, however, can also lead to bad habits, such as pulling on the reins, holding strongly, and not yielding when the horse responds properly. The goal in dressage is very light contact and self-carriage (see Glossary). The appropriate amount of contact on the reins will depend on the style of riding, the horse's level of training, and the proficiency of the rider's hands.

Pressure. Horses generally resist heavy steady pressure and respond favorably to light, intermittent pressure. That means if you are trying to get a horse to respond in his jaw and poll to your hands on the reins and you pull steadily and with great force, the horse will probably try to push into and out of that pressure, using the underside of his neck to resist your strong rein aid. If, on the other hand, you use a light squeeze-and-release action of the reins, the horse will tend to move in response to that action in his mouth: he will flex at the poll and flex his lower jaw away from the intermittent pressure on the bars of his mouth. Once a horse has responded this way, the rein pressure should become light and following.

The same pressure principle applies to the use of your seat and legs, and to the overall use of your body when you are on the ground. When riding, if you drive rigidly with your seat onto the horse's back he will probably tense and/or hollow his back in a form of resistance. But if you use your seat in a series of light, intermittent, collecting movements followed by a soft following seat, the horse is likely to respond favorably and begin moving with a more rounded topline.

If you are trying to get a horse to move his body away from hand pressure when you are on the ground and you lean your full weight into the horse, steadily and with all your might, the horse will probably push back into you. If, on the other hand, you use little taps with your fingertips on the horse's side, he will move away from the pressure. Similarly, when you are mounted, you will find that short, intermittent kicks with your lower legs are more effective than a steady grip.

Flexion and bending. Depending on who is talking, these words may be used synonymously, similarly, or differently. Flexion usually refers to vertical movement of the spine and limbs. It includes flexion of the joints of the horse's lower jaw, poll, neck at the withers, back, and croup as well as flexion of the knees, hocks, and stifles. Flexion is essential for impulsion,

Connection is also the relationship between the driving aids and the restraining aids.

collection, extension, and animation. Bending customarily refers to the left or right curving of the horse's spine. It describes the arc made by the horse's neck and back in a turn.

Rhythm. As previously mentioned, the regularity of a horse's steps in each gait is a very important component of riding. Each gait is like a simple musical piece written in its own specific time. Every horse plays that piece in his own particular tempo and with his own personal expression. Many horses have one or two gaits that do not have an even, precise rhythm. To be most effectively ridden and influenced by a rider, however, a horse must establish a very regular rhythm in all the gaits.

It is helpful when you are working on recognizing, establishing, and influencing a horse's rhythm that you train yourself to

Light rein contact with passive or giving hands and no driving aids results in this horse being above the bit. She uses the muscles of the underside of her neck to avoid taking contact with the bit.

Driving aids have been applied and the contact with the left rein has been increased. This results in left flexion and bending, yet it appears that it has also caused the horse to over-flex or come somewhat behind the vertical. The rider's wrists have broken the desirable straight line from the elbow through the hand to the bit.

have a very specific idea in your mind of what the horse's natural rhythm should be in each gait. You can count out loud or watch a metronome in your mind to monitor rhythm. Music can help you establish or maintain your sense of inner rhythm and, if appropriate for a particular gait, can be useful in developing a horse's rhythm as well. Continually refer to your horse's ideal rhythm patterns in your mind as you work. Otherwise, as your horse speeds up or slows down, he will tend to make you forget the target rhythm. Working on recognizing a horse's rhythm requires some basic information and a relaxed but very focused concentration.

The trot or jog. The trot or jog is usually a very regular gait with two distinct beats. The term trot refers to the gait as performed under English tack with a greater length of stride and impulsion than the western jog, which is shorter strided, and usually less energetic. The trot is the steadiest, most rhythmic gait of most horses; it feels more stable and precise than the walk or canter, and is therefore the gait that helps most riders develop their concept of rhythm. The horse's legs move in diagonal pairs like a metronome, clicking off an even 1-2 rhythm.

The right front and left hind legs rise and fall together and the left front and right hind legs work together. A rider can hear, feel, and see this rhythm. Without leaning over, glance at your horse's left shoulder and you will see a distinct change in the muscular shape and configuration of the shoulder as the left front leg reaches out, lands, and flexes. As the front shoulders alternate in their movement, their corresponding diagonal hind legs are also alternating.

The walk. The walk is a four-beat gait. When performed correctly, there is a very even rhythm between the feet as they land

and take off in the following order: left hind, left front, right hind, right front, left hind, and so on. This gives the rider a slightly side-to-side motion as well as a rear-to-front motion in the saddle. A horse that can really walk out gives what is called "a rein-swinging walk."

If a horse is in a hurry, though, he may have trotting on his mind and may not settle down into an even flat-footed walk. He may seem to be walking on the tiptoes of his hind hooves, which may make him jig (a cross between a walk and a trot) or pace (where the distinct four-beat pattern has been rushed and he is using his two right legs and his two left legs together, resulting in a lateral, pacey walk).

If a horse is very slow at the walk, the rider must try to encourage him to push off with his hind legs with greater energy. Before you try to change how energetically a horse is walking, however, you must first get in tune with his natural walking rhythm, even if it is too slow.

Sit with a relaxed seat and legs and let your body sway in time to the movement of the horse's back and hind legs. Do you find that in one moment your right leg swings against the horse's barrel while at the same time the left leg swings away from the barrel? And in the next moment the opposite occurs? Allow yourself to sway side-to-side exaggeratedly. Then, as your right leg does come in contact with the horse's right side, press with your right calf at that instant then immediately release so that you can press with your left calf when it swings against his body.

Continue this encouragement with your lower legs in time with the horse's rhythm and you will find that he will begin stepping more energetically and with longer strides. Then take a break. Stop applying the influencing aids and note that the horse may slow down. Then

Light, active rein contact accompanied by driving aids results in the horse taking a soft contact with the bit. He is now on the bit.

Light rein contact is suitable for warm-ups, riding young or inexperienced horses, and some types of western riding.

The western jog.

from rear to front, then during a moment of suspension the horse gathers his legs up underneath himself to get organized for the next set of leg movements. The rider seems to glide for a moment until the initiating hind lands and begins the cycle again.

Impulsion. The horse's locomotion originates in the hindquarters. As the discussion of the walk demonstrates, you affect the power source by getting in rhythm with the horse's hind legs. When a particular hind leg is landing, you can alter the energy with which it will next push off by adding seat and leg pressure on the same side. For example, if the left hind leg is getting ready to drive and you increase the pressure of your left calf, your horse will step forward with more energy with his left hind leg.

The time at which you apply your left leg will vary somewhat, according to what you are doing. The aid can be applied as the leg is landing, when it has landed, or just as it is getting ready to take off. Your instructor will help you identify when each particular leg is moving and the optimal time to affect its movement.

Collection. This is a state of balanced energy characteristic of a well-trained horse. Collection is characterized by a dropped croup, engaged hindquarters, flexed abdominals, an arched spine, an elevated head and neck, and a flexed poll. To achieve this configuration, a horse must be energized yet restrained. All through your early riding training you will hear your instructor tell you not to inflict opposing signals on your horse. In gross terms, riders are taught not to kick and pull at the same time. Yet as you and your horse become more advanced, you will begin introducing the concept of collection, achieved by applying seemingly contradictory aids — the

begin energizing him again. Do this until you can really feel a difference between the horse walking at his own rate and the horse walking at a more energetic rate as encouraged by your legs and seat. This is the beginning of your effective influence over the horse.

The canter or lope. The canter or lope is a three-beat gait that begins with one hind leg, then the diagonal pair, and ends with the leading foreleg. If the initiating hind leg is the left, the diagonal pair will consist of the right hind and the left front, and the leading foreleg will be the right front. The horse will be on the right lead. When observing a horse on the right lead from the side, his right legs will reach farther forward than his left legs will. The right hind will reach under his belly farther than the left hind, the right front will reach out in front of his body farther than the left front. If the horse is on the left lead, the situation reverses.

The canter has an alternating rolling and floating feeling to it. The energy rolls opposing forces of your driving and restraining aids.

Here is where optimal intensity and timing become critical. As you begin to practice the concepts of collection, you will find that if a horse is going to become confused, uncooperative, or fractious, it will be when you ask him to accept the confines of collection. Of course, approached properly, the lessons of collection should not create a problem, especially if you follow the cardinal rule of collection: never create more energy with your legs and seat than can be contained and shaped with your hands. If you do, the horse may try to escape by pushing through your rein or leg aids or may feel threatened by the restriction of the aids, and rear.

Lateral balance. This refers to the relationship between the movements of the left and right sides of the body. The goal of a rider is to not show sidedness, or a tendency to do things in a crooked or stiff manner. With some stiffness problems, flexibility exercises will help. Be sure your saddle and stirrups allow you to sit and ride evenly balanced.

If you or your horse have a structural asymmetry, it will be difficult to perform with lateral balance. In fact, in some cases it might not be in the best long-term interest to perform in lateral balance. For example, if you have one leg measurably longer than the other, yet force both legs to use equal stirrup lengths, you may be shifting your asymmetry up to your pelvis or spine. Or you may be twisting your trunk to compensate for the unevenness caused by the equal-length stirrup leathers. Get some professional advice on your condition if you find you have a persistent problem. You should never feel that you are forcing yourself into an unnatural "acceptable position."

The walk.

The lope.

In your early riding training, your instructor will tell you not to inflict opposing signals on your horse.

Lateral movements. These maneuvers contain some degree of sideways movement. The most extreme lateral movement is when the horse moves directly sideways, as in a sidepass or full pass. In between there are variations, such as western two-track, leg-yielding, and the half-pass. Although each of these movements has different standards for the angle and bend of the horse's body, neck, and throatlatch and the degree of collection and engagement, they do have one major characteristic in common: the horse moves sideways and forward at the same time. Lateral movements also include exercises such as the shoulder-in where the horse holds his body in a somewhat angled position off the track, a configuration that causes him to cross one hind leg over the other and one front leg over the other as he moves forward.

No Absolutes

Even though classic principles of riding, such as those taught in dressage and high-level western horsemanship, are successful for the majority of riders and the majority of horses, there are times when things just don't work. In those instances, rather than continue repeating aids or exercises or lessons that end in no progress or even a regression, a different tactic should be employed. That is where a good instructor will prove her worth. The training of each rider, to some extent, must be an individualized, tailormade program. A large share of the learning process will be similar for the majority of riders, yet following a program to the letter simply because of tradition does not always yield the best results.

As you become an advanced rider, the cooperative relationship between the driving aids and the restraining aids is what allows you to achieve impulsion and collection, as is exhibited in this canter pirouette.

In order to ride your horse in balance from side to side, your stirrups must be even. Here the right stirrup is longer.

Take the time to make the proper adjustment.

Have your instructor check your position from the front and the rear.

CHAPTER 10

SAFETY

If you follow safe horse handling and riding practices you will greatly minimize your chances of accidents. Most mishaps related to horses and riding can be traced to one of the following:

- Lack of understanding of horse body language; lack of experience handling horses; lack of ability; not having a way with horses

- Carelessness, lack of attention, and over-confidence

- Unsafe facilities

- Inadequate or improper training of the horse

- Inadequate or improper training and/or supervision of the rider

- Unsuitable horse

- Equipment failure

- Poor equipment fit

- Bad luck, such as horse spooks, slips, or falls

- Lack of planning for emergencies

- Loss of temper

A cardinal rule of horsemanship is "Don't over-mount." This can refer either to riding a horse beyond your capabilities or to attempting a maneuver that is too difficult for a particular horse. Riding at your level of competence on a conditioned mount using well-maintained equipment in safe facilities substantially diminishes the chance of accidents.

Commonly, when a horse is faced with a lesson that he is unable to accept, he reacts with an undesirable avoidance behavior. Rushed training can cause explosive reactions such as bucking, rearing, or running away.

Because of their size, strength, and quick reflexes, all horses are potentially dangerous. A horse's power is often underestimated and can surface unexpectedly, especially in an unfamiliar environment with a lot of commotion. At lessons and competitions, as well as at home, safe habits should be established and adhered to.

All equipment should be of the strongest materials and construction and should be periodically inspected for wear.

Stout gear is added insurance that a horse will not develop a bad habit. If a horse learns that he can break a piece of weak equipment and become free, he may develop a dangerous life-long habit of attempting to repeat such an escape.

Tack should therefore be well-stitched and constructed from durable materials. It should not be fatigued from sweat, weather, or dirt. Hardware should be of the highest quality affordable. Some tack is colorful or attractive, but may not be reliable under severe stress. Make dependability your number-one priority when choosing tack.

Care should be taken when passing other horses on the ground or when mounted. An exchange of kicks between two horses can happen suddenly and fatally injure a person or horse in the line of fire. Do not allow your horse to approach and sniff a strange horse. Often this is the prelude to squealing and more aggressive communication. Stay out of striking and kicking distance of both the front and hind legs.

Ride a safe horse.

Approaching a Horse

- Always speak to a horse as you approach him to alert him to your presence.

- Approach a horse at an angle, never directly from the front or rear.

- Touch a horse first by placing a hand on his shoulder or neck.

- Don't pat the end of a horse's nose; it often encourages him to nibble.

- Either walk around a horse well out of kicking range or stay very close, touching him firmly to let him know where you are. Never walk under or step over the tie rope.

Handling a Horse

- Know the horse you are handling.

- Let the horse know that you are firm but will treat him fairly.

- Control your temper.

- Don't surprise a horse. Let him know what you intend to do by talking to him and touching him.

- Learn simple means of restraint from a knowledgeable horseman and use them to control your horse when he becomes frightened or unruly.

- Stand near the shoulder rather than in front of the horse when clipping and braiding.

- Stand next to the hindquarters rather than directly behind a horse when working on the tail, whenever possible.

- Be calm and keep your balance.

- Do not make sudden movements or loud sudden noises.

- Do not drop tools or tack underfoot.

- Punish only at the instant of disobedience and do so without anger.

- Do not leave a halter on a loose horse as he may hook it on a post or a tree when rubbing or on his own hind shoe when scratching his head with his hoof. Many horses have died horrible deaths when people ignored this rule.

- Wear footgear and gloves to protect your feet and hands.

- When working in an enclosed space, always take time to plan an escape route in the event of an emergency.

Leading a Horse

- Make the horse walk beside you, not pulling ahead or lagging behind. Somewhere between the horse's head and shoulder is usually the safest position for you to stand or walk.

- When turning, turn the horse to the right (away from you) and walk around him rather than having him walk around you.

- Use an 8-10-foot lead rope. With your right hand, hold the lead 3 to 4 inches from the snap that attaches to the halter. With your left hand, hold the balance of the lead in a safe configuration such as a figure eight. Holding the lead in a coil rather than a figure eight in the left hand may cause your hand to become trapped in the tightened coil if the horse suddenly pulls. Use your right elbow in the horse's neck to keep the horse straight and to prevent him from crowding you.

- Work the horse from both the right and left sides so that he develops suppleness and obedience each way and does not become one-sided.

- If a horse resists, do not get in front and try to pull. Instead, stay in the proper position at the shoulder and urge the horse forward using a long whip held in your left hand.

- Never wrap a rope or strap around your hand, arm, or other part of your body. If a horse spooks suddenly and bolts, you may be unable to free yourself from him and could be hurt badly.

- Teach your horse patience when turning him loose. Do not let him bolt away. If he forms this bad habit, he may pull away before you have the halter fully removed and you could become entangled, kicked, or knocked to the ground.

Tying a Horse

- Know how to tie the quick-release (manger) knot without hesitation.

- Keep your fingers out of loops when tying knots.

- Be sure a horse is well accustomed to being tied in other ways before attempting to tie him in a cross-tie.

- Tie horses a safe distance from each other.

- Never unhalter a horse while the rope is still tied.

- Untie a horse and hold him temporarily with the lead around his throatlatch before removing the halter. This will help prevent him from developing the bad habit of pulling away.

- Never tie a horse with bridle reins. It would be far too easy for the reins to snap, which could damage the horse's mouth and encourage a bad pulling habit.

- Always tie at the level of the withers or higher and to a strong post or tie ring, not to a rail or a board that could be pulled loose.

Bridling and Saddling a Horse

- Wear protective headgear when handling the horse's head and ears until he is accustomed to you.

- Inspect your tack for signs of wear. Equipment failure is one of the leading causes of falls and injuries. Check girths, cinches, latigos, stirrup leathers, headstalls, buckles, and reins for signs of dryness, cracking, or metal corrosion.

- Pick out hooves and verify that the shoes are all there, that they are tight, and that the hooves have not overgrown the shoes.

- Estimate the bridle's adjustment and allow for a little extra room to ensure easy bridling. Make final adjustments after the horse is bridled.

- Check the horse and the saddle for foreign objects.

- Place the saddle and blanket forward of the withers and slide them back into position. Never slide it forward once it is on the horse because it will ruffle the hair and be uncomfortable for the horse.

- With a western saddle, fasten the front cinch first, then the back cinch, then the breast collar and accessories. Reverse the order when unsaddling. If a horse should spook with just a rear cinch fastened, the saddle could slip under his belly. This could cause him to buck, possibly injuring himself and you and damaging the saddle.

- Do not let the rear cinch be dangerously loose. Buckle it so that it is snug but not tight.

- With an English saddle, attach the breast collar and accessories to the girth, then buckle the girth. Reverse the order when unsaddling.

- If riding English, be sure the safety catches on the stirrup bars are in their open position. This will allow the stirrup leathers to slip away from the saddle in the event that you get a foot caught in the stirrup during a fall.

- Tighten cinches and girths gradually. Check several times before mounting and after riding a short while.

Riding a Horse

- Wear a protective helmet with full harness whenever possible. Protective headgear is a must when riding a young or new horse or when riding any horse over jumps.

- Never mount where there is low overhead clearance or projections.

- When mounting, maintain control of the horse through light contact with the reins.

- Confine your riding to an enclosed area until you know your horse.

- Remain calm if your horse is

frightened. Give him time to overcome his fear.

- Never fool around when handling horses.

- Don't rush past riders moving at slower gaits.

- Don't crowd other horses.

- Stay away from horses when your judgement and reflexes are affected by uncontrollable emotions, alcohol, other drugs, fatigue, or illness.

- Don't ride alone if at all possible. If you must ride alone, tell someone where you are going and when you expect to return.

- Take lessons on the longe line until you feel you can master trotting and cantering on your own. You may wish to continue to take longe lessons throughout your riding career to further develop your seat.

- Ride under supervision until you feel very confident in handling unusual situations.

- Don't ride around loose dogs, noisy equipment, or heavy traffic.

- Take your time. Don't take short cuts.

- Wear the proper attire for riding.

- Don't chew gum or eat candy while riding. You may choke on it or bite your tongue.

- Choose the footing you work your horse on wisely and learn to ride on various types of footing safely. Use an appropriate speed for muddy, snowy, or icy footing.

- Ride regularly and stay in good physical condition. Maintain a healthy weight, appropriate

strength, and a good level of endurance.

- Don't ride a horse that is too difficult for you to handle. You don't need to prove anything to anyone by trying to master a horse with a bad habit, a very sensitive horse, or one that is too highly trained for your level of riding.

- When riding with other horses, such as in a group lesson or a warm-up arena, take care not to crowd others. To pass a horse that is traveling in the same direction, move off the rail about 8 feet toward the center of the arena when you are several horse-lengths behind him, then move up and pass the horse. Do not return to the rail until you are several horse-lengths in front of the horse.

Don't ride alone if possible and be aware that loose dogs constitute a potential hazard.

Wear proper attire for riding.

Wear sturdy boots or shoes with heels.

These unsafe shoes will result in rub marks on the rider's ankles and could cause her foot to slip through the stirrup and be caught.

When passing a horse that is traveling in the opposite direction, use the customary way of passing practiced at the stable where you are riding. In the U.S. and Europe, riders most often pass left side to left side (just as in driving) with plenty of clearance between horses.

In a horse show class, the slower horses hug the rail and the horses that move more forward or more quickly must pass on the inside. The opposite occurs in a warm-up ring at a show, however, and in many commercial arenas where different types of riders work together. Often the riders that are working more slowly stay toward the inside of the arena, and those working at the canter or lope, for example, have the rail. This, however, poses the problem of how the faster riders pass the slower riders. Since it is dangerous to pass between the wall and another horse working on the wall, it is the custom for the rider approaching from the rear to call out "rail" so the slower riders can let her pass safely along the rail.

CHAPTER 11

YOUR MENTORS

As you learn to ride, you will become aware of the importance of your two mentors: your horse and your instructor. Choose them wisely, treat them with respect, and your learning experience will be enriched.

THE HORSE

It is important to separate the goal of riding a horse from the goal of training a horse. To train a horse, you must first become a good rider. To concentrate on your riding, however, you need the help of a well-trained, experienced horse. As you progress, the type of horse that best suits your needs will change. At the beginning, therefore, it may be best to ride your instructor's school horses. That will give you experience and time in the saddle, which will help you decide what type of riding you wish to pursue. Then you can consider purchasing a horse that is appropriate for your riding-skill level.

A novice learns the basics most easily on a patient, well-trained, experienced school horse — essentially a horse that knows more about riding than the rider does. In a similar vein, the advanced rider benefits from riding a more advanced horse to learn the components of the more sophisticated maneuvers. While it is true that a person can still improve her riding on a horse that she is also training, progress will be much more rapid on a solid, steady, well-trained horse of the appropriate level.

Horses of all breeds and types are suitable for beginning riders. Yet it is easier and more successful to use a school horse that is naturally suited for the particular type of riding being taught.

The Good School Horse

Most really good school horses are between the ages of eight and twenty or older. An aged horse (over five) that has

An experienced, well-trained horse and a qualified instructor will greatly accelerate your learning process.

not yet been trained or has been ridden incorrectly can be trained to be a school horse, but it is more difficult to convince him to change his physical and mental ways of doing things. A horse that is accustomed to following his own natural inclinations, such as unbalanced, meandering movement in the pasture, might be quite resistant when asked to straighten up and move correctly for lessons. A very old horse that is gentle but set in his ways and has never been schooled properly or thoroughly would probably not be a good school horse.

Using a young horse often proves difficult for lessons as a young horse's responses are still inconsistent; a young horse has not experienced enough of the world to react calmly to unusual situations; and a young horse is not adequately developed physically to accept the imbalances of a learning rider.

Using a horse that has previously been used for another type of riding can sometimes work, but often the crossover is too confusing for the horse. A western horse may have been allowed to lope rather flat and on the forehand for years. If he is then used for English lessons, when he is asked to accept contact and move in longer strides in the canter he may show confusion, resistance, and a reluctance to change his old habits. In addition, because of his conformation he may have difficulty performing what is being asked of him. Conversely, a horse used for fox hunting that has been rewarded for years for moving in an extended, energetic frame might have a hard time adapting to the very slow pace of western riding.

The best school horses are usually geldings. Stallions are rarely appropriate for learning riders, while mares' suitability varies according to the phase of their heat cycles. Mares may be more useful in the winter months when estrus is not occurring. During the spring, summer, and fall breeding season, however, mares that exhibit strong heat periods may be frustrating to ride and inappropriate for a learning rider. Some mares, due to high levels of estrogen, become preoccupied with staying near and retaining contact with other horses, not necessarily stallions. Sometimes their behavior is quite exaggerated: stiffness, unwillingness to pay attention to the rider, silliness, whinnying, squealing, short and choppy strides resulting from excitement and tension, swerving toward the gate of an arena or the pen of a preferred associate, and so on.

The school horse must have an exemplary temperament. He must be patient, willing, cooperative, and alert, yet calm. He must be physically responsive to the aids and balanced and rhythmic in all of his gaits. A very sensitive, thin-skinned, hot-blooded horse may be desirable for a professional in competition, but may be disastrous for a learning rider, reacting to every bump of the leg or hand as a signal

The school horse for a beginner must have an exemplary temperament: patient, cooperative, and calm.

to move his body one way or the other. The beginner usually learns better on a duller, more cold-blooded horse who will tolerate the mistakes a rider makes when discovering balance and rhythm. Such a horse tends to go on steadily despite the awkward movements of a rider.

For the intermediate rider, however, a dull horse's responses may be frustratingly slow and inaccurate, even if the rider is applying the aids correctly. Therefore, a moderately sensitive horse is ideal for the intermediate rider.

The advanced rider requires a sensitive, talented horse. This will allow optimal development of precision and timing in the use of the aids. The rider can improve her riding techniques more quickly if the horse's level of training is more advanced than hers.

Conformation

Certain points of a horse's conformation can make it easier or more difficult for a rider to improve. If a horse is lower at the withers than at the hip, for example, he will travel on the forehand. This makes it much more difficult for the rider to sit in balance and to keep her legs underneath her seat. As she tries to compensate for the downhill slope of the horse by leaning her upper body back, either her lower back becomes hollow or her seat slides forward, causing her legs to shift forward too.

If a horse's withers are higher than his hips, on the other hand, he will naturally tend to carry more weight on his hindquarters. The more weight a horse carries with his hindquarters, the more easily the rider can balance on the center of the seat of the saddle. This allows her shoulders to stay directly over her hips and her legs to come under her body.

If a horse has a moderate spring to his ribs the rider's legs can find a comfortable, natural position. Horses with very round barrels force the rider's legs to spread widely. Sometimes only the top of the legs have contact with the horse's back; the lower legs are pointed off the horse's body

The advanced rider's horse should be sensitive, responsive, and alert.

in an A-frame configuration. Conversely, horses with very flat ribs or a pinched heart girth make natural leg contact difficult: when a rider's legs hang naturally on such a horse, they will not touch his sides. In order to make contact and hold her lower legs on the horse, the rider will have to exert constant thigh muscle contractions. This would result in muscle tension, then fatigue, and ultimately in a loose seat.

Bad Habits

Certain habits are unacceptable in a school horse. Understandably, bucking, rearing, shying, running away, balking, biting, and kicking should not be tolerated in any horse, especially a school horse, but a school horse must be even more exemplary. He must not resist by running through the aids, speeding up, or coming above the bit.

A horse is not an opponent that you must force into submission but a partner with whom you must establish the terms of a working relationship. You must direct things in a consistent and fair manner so that energy flows harmoniously from you through the horse and then back to you, just as it does with you and your dancing partner. Remember, however, when you dance with your horse, *you* should be leading.

During lessons, try to not be defensive about your own horse. Some horses that don't at first appear desirable to your instructor begin performing well after three or four lessons. Others that initially seem to be the right type might not have the qualities necessary to help you learn. It is a surprisingly common and very serious error to think that a horse that is not good enough for competition would make a good lesson horse. This is similar to the

A school horse should have good habits. He should never resist by running through the aids.

classic false conclusion: "Since this mare is unsound (or unreliable or uncomfortable) when I ride her, I'll make her a broodmare." Although a school horse does not have to be beautiful and fancy, it does need to be sound, relaxed, cooperative, and rideable. Some horses may teach a rider ten things wrong before allowing her to learn one thing right. If a horse makes it almost impossible for you to learn, seek the help and advice of a professional in selling the horse and selecting a more appropriate one.

A horse is not an opponent but a partner.

LESSONS

Riders of all levels benefit from instruction. Riders in the following categories should particularly consider taking regular instruction:

- Children who know their left from their right and are old enough to concentrate and secure enough to interact and have fun. Lessons can start about the same time the child begins school at about six years of age or when the child weighs at least 50 pounds and is 48 inches tall.

- Riders of any age at the beginning stages.

- Riders with a persistent problem that causes them physical discomfort when riding.

- Riders that are fearful of handling or riding horses.

- Riders at the intermediate stage who have professional or competitive goals.

- Riders at the advanced level who are interested in developing finesse and refinement.

- All instructors no matter what their level of accomplishment.

THE INSTRUCTOR

There is no substitute for a good instructor, yet a poor instructor is worse than none at all. It is essential that you learn correctly and that your lessons progress in such a manner that you maintain a good opinion of yourself, your horse, and your work.

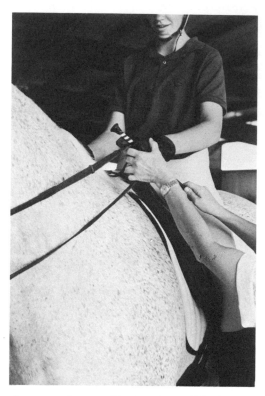

An experienced instructor will help you learn new techniques correctly, such as the Fillis method of holding the reins.

Characteristics of a Good Instructor

- She inspires respect and confidence as a teacher and a person.

- She respects the student's goal of competition or personal development even if it is different from her own.

- She is an experienced rider.

- She is a knowledgeable and respected horse person.

- She communicates in clear, easy to understand language.

- She is assertive rather than aggressive.

- She is effective in her teaching, not just efficient.

- She has appropriate relationships with her students. (This can vary among students: friend, coach, colleague.)

- She is concerned with safety. She has recently taken a first aid course.

- She knows there are many types of students and knows how to deal with them all effectively.

- She understands the physical progression of a rider's development as an athlete.

- She understands the mental processes involved in learning an athletic skill.

- She is orderly, organized, and punctual.

- She is genuinely interested in teaching and does not over book herself so that she becomes stressed and loses her temper or teaches poorly.

An excellent trainer may be able to teach you skills but if in the process she undermines your self-confidence, it is not worth the investment of your time, money, and trust.

The question is often debated as to whether a parent can be a child's riding instructor or whether a husband can teach a wife or vice versa. Rarely do either of these arrangements work to everyone's advantage over the long term. My advice is to seek the services of a qualified professional. It will be worth your investment.

Lessons should be enjoyable and should increase a student's confidence and preserve her self-esteem.

- She is warm but not gushy.

- She knows how to handle emotional and stress-related problems.

- She is confident but not aloof.

- She has a healthy ego that allows her to conduct the lessons with the students' interests in mind, not her own.

- She sincerely wants her students to surpass her own skill at riding.

- She knows the effective balance of praise and constructive criticism that works for each student.

- She looks to improve her own riding, teaching techniques, and training experience.

- She accepts part of the responsibility when a student has problems in a

lesson and knows what to do to get a lesson back on track.

- She has good psychological stamina and emotional maturity which allows her to remain steady and calm.

Choosing Your Instructor

What traits *you* will emphasize when you choose your instructor will depend on your goals. If you are trying to develop your awareness as you ride, you will want an instructor with a keen eye and the ability to give you accurate feedback. You might think of your instructor as an experienced, talking mirror. In order to give you top-notch feedback, your instructor should ideally be an excellent rider, trainer, and observer; someone who can recognize when things are going right and can immediately tell you so that you can register the feeling; someone who knows when things are headed in the wrong direction and can tell you in clear terms how to fix it.

Succinct, accurate verbalization is the ticket during the actual lesson. Long talks or discussions usually do not facilitate developing awareness and riding skills. In fact, many instructors talk too much during the lessons. They either present too much cerebral information when a rider is trying to develop motor skills or they keep the air filled constantly with the sound of their voice so that when there is one tiny silence, it makes the rider anxious that something is wrong.

As you consider the various instructors in your area, talk with those you are considering. Ask their students how long they have studied with the instructor and how they would describe their progress. Observe for yourself the instructor's

A qualified instructor can help you learn to communicate with your horse.

style of teaching, manner, and rapport with students.

Be sure you look for an instructor that is appropriate for you. For this, you must make an honest personal assessment of your goals and capabilities. Some trainers are excellent with beginning riders but do not have the proficiency to take them further. Others do not have the patience to work with any but very advanced riders and horses. An instructor with a famous name may not have the time or interest required by the developing rider. So if a well-respected, accomplished instructor is willing to work with you, by all means take advantage of the opportunity.

Once you are fairly certain you wish to study with a particular instructor, again observe her at work so that you can become familiar with the style of speaking and the phrases and commands that are used in the lessons. Take note of the types of exercises she asks her students to perform and be sure you know when the work is being done correctly.

When you accept a particular instructor, you also accept her system of training. Hopefully, your instructor follows a well-respected system yet has the experience to modify it when necessary. Systems, by nature, are rather rigid, filled with specific rules and standards and a prescribed way of doing things. In that way, they are most effective when getting from point A to point B. Skipping from one system to another, on the other hand, and using bits and pieces from here and there can destroy the continuity of learning how to ride.

Give your instructor's methods a chance to work. If you are convinced mentally that they will not work, they will not work for you physically. Be sure you understand the overall concepts of your instructor's system; otherwise it can lead to misunderstanding. Misinterpreting your instructor's underlying philosophies can be counterproductive; if you are confused, it will eat away at your confidence in your instructor. With such a mind-set, you will find yourself applying half-hearted aids, *looking* for failure in your horse. The top-notch instructor inspires confidence in her students.

Getting the Most from Lessons

Before each lesson, and especially before the first, when you might be a bit apprehensive, take the time to relieve your and your horse's anxiety. The suspense of not knowing how you and your new instructor will work together is normal. If you are relaxed, you will have a better chance of hearing her comments and being able to do what she says. Use some relaxation techniques, such as your favorite stretching and breathing exercises, to dissipate muscle tension and get rid of the butterflies in your stomach. Check ahead of time to see if your instructor prefers you to bring the horse into the lesson warmed-up or cold.

Contrary to what students think, what most often blocks progress is not a poor horse or a poor rider body or a poor knowledge of theory and techniques, but a poor mind-set. A positive and receptive attitude is of paramount importance in your lessons. Keep an open mind and take what your instructor is trying to give you.

It is puzzling why a student with a doubting, negative outlook bothers to take lessons in the first place. Yet a great number of students are this way, often without realizing it. It seems logical that if you are investing your time and money, you should be ready to learn. Listen to your instructor's advice and apply it open-mindedly and diligently during and between lessons. Often, results are not possible in just a few sessions. Switching instructors or methods just when you are getting started may not be wise unless you find you are regressing.

Lesson progression. Even though every horse and rider present a different set of problems for the instructor to solve, the beginning lessons are usually standardized. For example, all new students may be directed to carry the hands 4 inches above the withers and close together. Later, variations may be added, such as lifting one hand or carrying one low, but initially it is necessary to standardize the rider's aids.

At any one time, an instructor has several things that she can work on with each horse and rider and may often select something to add a little variety to the lessons for herself. For example, even though there are few riders who couldn't use further help on seat, an instructor may decide to bypass rider position temporarily

and concentrate instead on getting the horse to round into the bridle on a downward transition from the trot to the halt. Just because the instructor did not choose to school the rider on her seat does not mean it did not need work. Especially with a rider that is not advanced, there is so much to do at first — rider position, control of the gaits, speed within the gaits, bending in corners and on circles, and so on — that working on one aspect is about the same as working on another. Only one thing can be emphasized at a time, so slow progress is made in a variety of areas until things "start coming together." The rider that doesn't understand this might read between the lines and think, "Because she didn't work on my position, I must be doing OK." Don't try to second-guess your instructor's intentions.

The overall lesson progression that is followed for most riders is:

- the development of awareness in the horse and the rider
- the development of the seat
- the establishment of energy and communication between horse and rider
- the creation of desired responses in the horse
- the development of accuracy

Sometimes an instructor will overlook an error in one area to enable you to achieve success in another. For example, if you have ineffective aids, such as poor leg contact, making the horse move lazily, the instructor may tell you to use the spur or whip. This might create a quicker rhythm than would eventually be desired but will allow you to feel the difference in energy. Tempo might be compromised temporarily so that a different reaction can be established with the horse and recognized by you. The instructor selects what would be best to work on and you must trust her judgment and experience.

Changing old habits. If you come from a different type of riding, you might be holding on to some of the principles from that discipline — unknowingly or in an attempt to feel comfortable. You must become aware of your old habits in order to change them, or else they will hold you back. Understanding the goals and purposes of your new type of riding, as well as your previous style, will help you change more easily. Most importantly, you must shed the constraints of the familiar and be willing to try a new way of doing things. It is a false security to hold onto something just because it is familiar.

Some students repeat the same pattern over and over because they are unconsciously more comfortable knowing the result of an old technique — even if it is not effective — than with a new one. But often it is just that the student does not know what the result should be or should feel like. Usually it is easier to learn something if you use a totally new procedure than if you try to modify your existing system. But you must be willing to risk in order to win.

Sometimes an instructor will ask you to work your normal routine so she can see your and your horse's skill level and tendencies and evaluate the work you have been doing on your own. Although no instructor wants to make discouraging statements at first about the work, she must at the same time be honest.

If, for example, a rider is applying incorrect aids, then the instructor has an obligation to point it out, even if the student is an accomplished competition winner. If an instructor has substantiated

expertise, her observations and constructive suggestions should be respected.

It is best if you do not take comments about yourself or your horse personally. Constructive criticism during a riding lesson is considered an essential part of learning. The criticism should be delivered tactfully, however, never with the intention of belittling you or undermining your confidence. If you are sincere and your instructor ridicules you, for whatever reason, discontinue the lessons and find another instructor.

When receiving warranted, constructive criticism and advice, don't be argumentative. There is a fine line, however, between making excuses and conveying essential information to your instructor about your limitations. If you have a physical problem, state it beforehand so you are not overworked. If you claim a limitation just to get out of working hard, you are losing an opportunity to advance.

Riding is generally a mirror of your character. Your nature will show up in your lessons and in your daily work. Even with proper instruction, if a person has a tendency to be too careful or too aggressive it will show up in her techniques or in her responses to the instructor's comments. If a person does not try to get more balanced within herself, her riding will not change. Something in the person must change or the same tendencies will repeat themselves over and over.

Sometimes a student will say, "This horse hates this exercise." First of all, it is an anthropomorphic trap to assign human emotions to a horse. In actuality, it is the rider who dislikes the work. Once a rider learns the effective use of the aids, the horse's expression changes and both the horse and rider begin to like what was

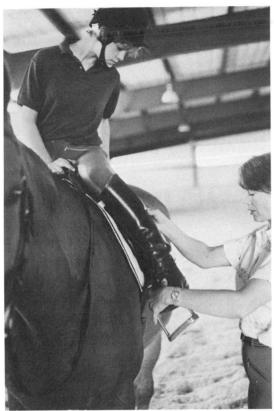

An instructor should communicate in a clear, easy-to-understand way.

previously either a frustrating or a boring exercise.

In order to get the most out of your lessons, learn to interact with your instructor in the way acceptable to her. Many instructors prefer that you do not talk when you ride. If you are asked, "Do you understand?" or "Do you have any questions?", listen to how you reply. If you say "I think so" or "I don't think so," you are probably speaking with a questioning inflection. Be decisive in your thoughts and words. If you don't understand, say so.

Sometimes because of the noise of your breathing or heartbeat, the squeak of your saddle, the sound of your horse's hooves, or the wind, you cannot hear your instructor. If that is the case, be sure to tell the instructor rather than straining to hear

and guessing what you are being asked to do. Your instructor should have the proper equipment to amplify her voice if it is not loud enough for you to hear.

As you receive feedback from your instructor, you will be subconsciously adding to, deleting, modifying, and clarifying the visualization images in your mental checklist. If you are not sure exactly what you are doing correctly when your instructor says "Good," then ask. The same applies, of course, when the instructor calls out for you to change what you are doing. Be sure you understand how much you should move your leg back and when it is just right, so you can register that in your mind-set.

Positive vs. negative lessons. The overall feeling you get during your lesson can be one of positive momentum or a disastrous cycle of mistakes. During a positive lesson you continue building on your successes throughout the lesson, adding new techniques or exercises that build to a greater performance from you and your horse. Your instructor will recognize the optimal time for short breaks. She will also know the best time for the lesson to end on a successful note so that you do not destroy what you have gained by going beyond your energy or skill level.

An interesting phenomenon occurs with many riders during lessons. When the instructor praises the rider, the rider experiences such a sense of elation that there is a relaxation of the aids and things fall apart. During a lesson you need to stay on Earth, concentrating on riding every step even though your instructor's approval may understandably put you on cloud nine.

The lesson with negative momentum may start with a major error followed by a strong remark by your instructor. This arouses a fear in some riders, a fear much greater than falling off — the fear of failing. A riding lesson can be an excellent opportunity for failure. Each person has certain circumstances that make them anxious: outside distractions, the opinions of onlookers, a horse behaving less than perfectly, and so on. If you let these outside influences affect you, you are setting yourself up for failure. You become flustered to the point that you cannot even perform elementary exercises correctly. This happens because external factors have caused you to focus on preserving your self esteem rather than on performing what the instructor is trying to teach you.

When a lesson disintegrates into frustration, the instructor has basically four choices: to take a break; to end the lesson; to push you through the trouble spot if you are mentally and physically able; or to go to less demanding work to reestablish a line of communication. Lesson anxiety usually diminishes once the rider is in good physical shape, the horse is better trained, and both are used to the routine of the instructor and the lessons.

It helps to have a positive attitude but realistic expectations about your progress. While a good instructor can often spot the very thing a horse and rider need to work on in the first lessons and make a significant break-through, this will not happen every lesson. There are exciting milestones but also seemingly endless plateaus where progress may appear to be at a standstill. Here again the instructor's judgment must be trusted. If you have doubts about the qualifications or effectiveness of your instructor, seek the opinion of well-respected professionals in the area. You may find that your performance is poor in part due to your teacher's low expectations of your success. Sports and educational psychologists have found that athletes and students perform to their teachers'

expectations of them. If you progress consistently in your work at home and regress during lessons, it may be your instructor's opinion of your ability that is holding you back.

Between Lessons

What should you do between lessons? While there are a lot of attractive peripheral activities associated with riding, they should be kept in perspective. Don't go overboard with excessive grooming, worrying about theory, or fussing with tack. The main activity in the development of the rider is riding. Between lessons, spend as much time as possible in the saddle. This cannot be overemphasized.

Some people are generally not satisfied until they know the reason behind every single thing they are asked to do — they are hung up on theory. But in riding, some things cannot be explained, they must be developed. More important than finding the rationale behind every single movement is learning to recognize when your seat is deep, your hands are soft and yielding, and your lower leg is acting correctly.

In between lessons, you can rehearse a particular movement until you get the sequence of aids established in your mind. For example, with the leg yield or two-track, there is a necessary coordination and timing of the aids. If you review the order of the aids in your mind until they become an automatic response, the intellectual aspect of what you are doing will not predominate when you are riding. This will allow you to abandon yourself to the development of feeling during your daily rides and during your lessons.

Between lessons, it is also useful to think about the main point that your instructor was trying to make in the lessons. Often this is an abstract idea such as to be less timid or to ride the horse more energetically. Although there are specific techniques that might help you understand these nebulous ideas, the systematic combination of aids and exercises is not more important than developing a state of mind suggested by the instructor. If a student hears that she is not assertive enough, she should alter her way of thinking so that she can do something about it. The rider must be aware of the overall goal, the big picture, rather than getting hung up on small details. Having your lessons videotaped will allow you to review the good moments and the problems in order to see and understand what the instructor was trying to get across.

Reading suggested reference materials in preparation for your lesson will save you time and money in the long run. Ask your instructor for a list of books that would be helpful for you to read. Keep a notebook or file box listing various exercises and techniques that you have learned, so that when you come to a trouble spot or pass through the same stage with another horse you can refer to your notes.

Working with a Clinician

Many of the principles that apply to working with a regular instructor are also appropriate when you ride in a clinic. There is one significant difference, however, between how your regular instructor and a clinician will work with you. Your instructor, who knows you well and looks to preserve a long-term relationship with you, may tend to work with you more subjectively. A clinician merely analyzes your current problems, prescribes corrections, and delivers all in a direct and objective manner.

A clinician doesn't have the opportunity to get to know every horse and rider well and it is difficult to know what is safe to give each rider to work on. That's why it is best to have several lessons with the clinician so that there is a chance for things to come together before the clinic is over. The student can look back at the first day and understand why the instructor did things in a certain way.

If the techniques you learn in a clinic bring good results, they should be incorporated into your regular training program and you should work with the clinician again when possible. If you have no success with the techniques, you may not be ready to go to clinics. You may benefit more from working with a regular instructor. If there is not one available in your area, organize some other students and arrange for an instructor to come on a regular basis.

Even if you are an experienced rider, be careful of going to a large number of clinics that promote different approaches. The contrast in theories and techniques can seem contradictory and confusing. While one clinician may emphasize rider position as the key to getting a horse round and balanced, another instructor will spend the majority of the time focusing on the horse itself, through flexing and bending exercises. And even among those clinicians who zero in on the horse rather than the rider, each one approaches the lessons with slightly different exercises and techniques. While an advanced rider may be able to see the similarity in the aim of all of these methods, the less experienced rider might easily become perplexed.

CHAPTER 12

THE AIDS

Learning to feel and use the aids is the most important aspect of riding. Rather than gathering a multitude of exercises, use the basic movements of a horse as vehicles for your development. Throughout your progress as a rider, you will shift your focus from the parts to the whole and back again. As a beginning rider you should work on overall skills. Once you have achieved a degree of confidence, balance, and rhythm, you can focus on the individual aids. When an elementary competence has been attained, you can approach riding once again as a whole. This shift of focus from the parts to the whole will continue as you develop a mastery of the aids.

Understanding the composite effect of aids is more important than memorizing "sets of aids" or "cues" to get a horse to perform various manouvers. Approach riding as a series of coordinated body movements rather than memorizing what buttons to push. For example, instead of thinking, "Canter to trot," think "Shoulders back, sit deep, squeeze with outside hand"

and you will find that your horse has made the transition from a canter to a trot. This way you feel how your various body movements create various reactions in your horse. And in this way, you can more accurately focus on which of your aids are working and which need development.

Then, to gain a full appreciation of the aids, rather than focus on each aid individually and intently as you ride, transfer your awareness of riding from a part to the whole. Diffuse your focus from a single segment, such as the rein aids for leg yield, to a larger context, such as the feeling of flowing forward and sideways at the same time.

To do this you must separate the means (the aids) from the goal (the leg yield). This is the basis for the cybernetics theory of achievement. With cybernetics, you must first identify and locate your target. Second, you must create and sustain the necessary energy to reach the target. And third, you must develop a feedback mechanism and pay attention to what it tells you. You will be more likely to achieve your goal if you focus on the positive feedback — your successes as you ride.

First, focus on the end goal (the leg yield) rather than on the aids. Then energize your body and the horse's so that you create forward movement. Then, as you apply the aids, remain open and receptive to feedback from the horse via your body. In spite of the continuing stream of information you get from the horse as you ride, you will usually be able to identify certain things that you do with your body that are successful and should be continued. Build on these effective uses of your aids. Make any adjustments to your body position gradually so you don't disrupt your positive gains and you can stay on target and reach your goal.

THE NATURAL AIDS

You influence a horse through your use of the natural aids and, when necessary, some artificial aids. Your natural aids are your mind, seat, upper body (weight), legs, hands, and voice. The mind, a powerful aid, was discussed in the first section. The voice is a means of communication among humans, not as effective in riding training as it is for ground training. Artificial aids are items such as spurs, whips, and martingales that accent or augment your natural aids.

You cannot haphazardly apply your aids and influences with any hope for success. Think about the following overall rules regarding the application of the aids before you learn more about their specific effects.

- Your aids should be fair and not contradictory, humanely applied

and understandable by the horse in equine terms. There should be no conflicting aids that would make a horse unable to respond properly to one aid without being punished by another.

- Your aids should be appropriate. You could train a horse to stop when you blew a whistle that you held in your mouth all the time you rode. But would that be convenient or appropriate? It is most appropriate to use an aid that is related to a horse's natural reflexes. A horse's natural reflex in response to a whistle would be to lurch forward. Using a deep seat to burden the croup is a more logical and appropriate aid for stopping a horse's movement. Following classic riding guidelines is the best bet.

- Your aids should be clear and direct. Since a horse does not reason, you must make your signals to him a direct line of communication, not something he must figure out. Know the horse's level of training so that your aids will be delivered at an appropriate level.

- Your aids should be consistent. Each time you want the horse to perform a particular maneuver, you should ask him in the same way he has been asked in the past.

- Your aids should be precise. They need to be applied at the right moment so that you make it easy for your horse to do the right thing. Your sense of timing will improve with practice, so it is best not to attempt a very complex maneuver until you have the ability to coordinate all of the aids for it.

■ Your aids should be applied in the appropriate location. You must know horse reflexes and how a horse will react to stimuli applied in various regions of his body. If you are trying to get a horse to move his right shoulder to the left, it would make no sense to use your right leg behind the girth; this might actually make him turn on his center and move his shoulders to the right. Similarly, if you are carrying a whip in your left hand in front of your leg as you ask the horse to move his right shoulder to the left, can you see what a conflict the horse might experience?

■ Your aids should be of the optimal intensity. They should be applied strongly enough to be effective yet not too strongly or they might frighten the horse or develop resistance in him. An instructor can be helpful when determining how much to squeeze the reins or bump the horse's side with your leg in order to get the desired results.

One goal of riding is to get a horse to respond to lighter and lighter aids. Yet often in lessons the instructor is heard calling, "More leg" or "More forward"! This does not mean to apply more steady *force* with an aid, but rather to use it more effectively. An abrupt, well-placed kick with the lower leg often gets an otherwise dull horse to respond.

Use an aid lightly and increase as needed. If a horse doesn't respond to a natural aid, you can reinforce it with an artificial aid. Whether you use natural or artificial aids, use them with only as much intensity as you can control. For example: You want your horse to move away from your right leg. You haul off and kick your horse in the right side with the heel of your boot, perhaps with a spur, and maybe

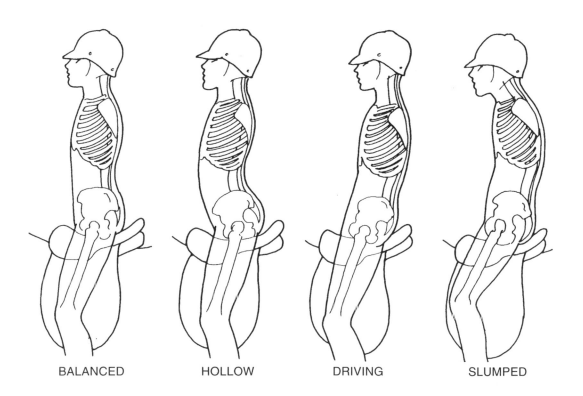

BALANCED   HOLLOW   DRIVING   SLUMPED

Don't lean your upper body forward thinking you will encourage the horse to move forward. This just causes a hollow back and loosens your seat.

complicated rein aids but instead is just lightly guided into position.

Your aids should always be followed soon (within a second) by a reward or release for the horse when he complies with your wishes. If you ask a horse to slow down by deepening your seat and closing your hands on the reins, and he slows down but you continue to hold him with your seat and hands, he has not been rewarded for doing as you asked. If you make this error several times in a row, the horse will lose the incentive to slow down for you in the future. Why bother? If you relax the aids when your horse complies, however, not only will he be willing to repeat that behavior for you in the future, but he will also look forward to it.

even with the addition of a whip. Your horse swerves exaggeratedly to the left and throws up his head. Now you have two things to correct. You have created a worse situation. You not only need to apply the aids with an appropriate intensity, but you must also be ready with counter-balancing aids (in this case, the left rein and leg) to contain the horse's response and help shape it into a desired form.

The more responsive a horse is to your seat and leg, the lighter he will automatically become in the bridle, because the seat and the leg shape the horse's body and give it its degree of impulsion and flexion. If the aids are applied correctly, the horse will carry his head in a naturally pleasing configuration that does not have to be altered with

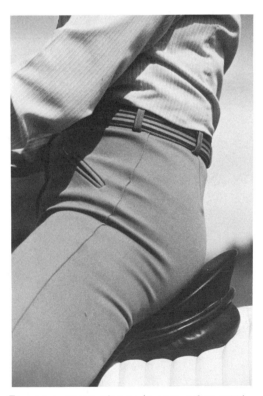

To encourage a horse to move forward, sit deep, contract your abdominals, flatten your lower back and use a driving seat.

The Seat

The foundation of good riding begins with a balanced, symmetric, "following" seat: the rider's pelvis must follow the motion of the horse's hindquarters in order to influence him effectively. In order to attain such a seat, you must be sitting on a saddle that allows it to happen. To help find your balanced seat while at a halt, keep your legs underneath you, stand straight up in the stirrups, and then lower your body into the saddle, letting your weight drop into your heels. It will be as if you are standing on the ground with your knees bent. Register this feeling and refer to it routinely as you work your horse in motion.

Every type of riding has its own standards and style of elegance, and a rider's overall position will vary accordingly. A basic, balanced seat, however, is essentially the same for bareback riding, western riding, and dressage, three types of riding that many riders enjoy. Variations of the basic, balanced seat include hunt seat, saddle seat, the endurance racer's seat, the cutting horse rider's seat, and so on.

Overall position of the basic, balanced seat when viewed from the front and rear: There is even weight on both seat bones; the legs lie softly along the horse's sides; the stirrups are even; the shoulders are level and square; the head faces forward; the neck is straight.

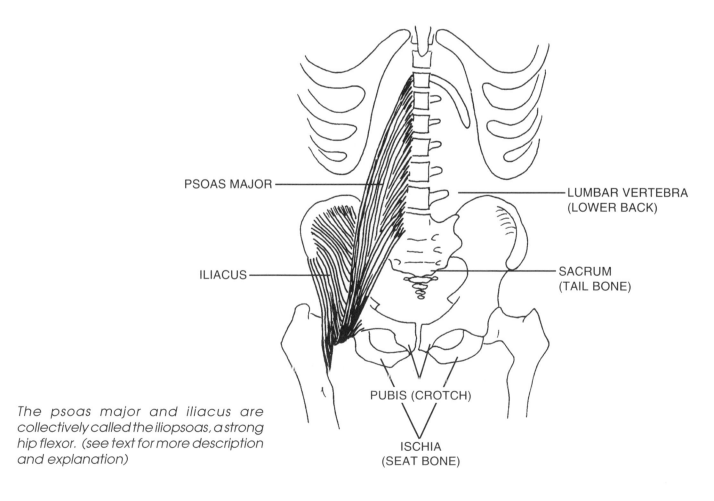

PSOAS MAJOR

ILIACUS

LUMBAR VERTEBRA (LOWER BACK)

SACRUM (TAIL BONE)

PUBIS (CROTCH)

ISCHIA (SEAT BONE)

The psoas major and iliacus are collectively called the iliopsoas, a strong hip flexor. (see text for more description and explanation)

Overall position of the basic, balanced seat when viewed from the side: There should be a straight line from the ear through the shoulder joint through the hip joint to the back of the heel. The rider's pelvis is in a position where the lower back is flat and by the action of the abdominal muscles the seat bones can be pushed downward and forward.

Common problems. One common seat problem involves a rider who perches forward in the saddle, rolling onto the crotch; the seat bones no longer have contact with the saddle. Such a rider, most typically female, tends to have a hollow back, which destroys balance and prevents her delivering effective aids with her seat. Because she is ahead of the motion of the horse she can easily be popped off over the horse's head or shoulder if he should suddenly stop.

This rider needs to tighten the abdominal group (especially the iliopsoas muscle) which will flatten her lower back and rock her seat bones under her. Additionally it will open her hip joints and allow her thigh to lower alongside the horse.

Another tendency among male riders is to sit in the saddle as if it were an old stuffed chair. In this case, the rider's buttock muscles are rolled back against the cantle. The rider assumes either the easy-chair position or the "braced-for-action" position. In the easy-chair position, the thighs are raised and held nearly horizontally, which raises the lower leg and brings the heels up. Such a rider is a passenger, definitely behind the motion, and might be left in a cloud of dust should the horse take off. In the braced position, the legs are held stiffly forward with the feet braced in the stirrups. This rider cannot feel the horse's back or follow his motion and there is a great chance that she

Perching forward on your crotch and bringing your lower leg back results in a hollow back and a loose seat.

Sitting back on the cantle and shoving your feet forward results in a rigid, braced position.

will experience stress injuries due to the rigid position.

This rider needs to rock forward more on the pubic bone. The military solution for such slouching posture is "Shoulders back, chest up and out, belly forward," which results in more weight on the pubic bone but also a tendency for a hollow back. Because a hollow back is unlikely in male riders, this sort of correction may be suitable for them. For a rider with female conformation, however, this could prove counterproductive and damaging to the spine.

Many riders will collapse their bodies to some degree during their riding training. The term "collapsed hip" can cause some confusion. If collapse is thought of as a breaking down (such as a roof collapsing toward the ground), then a "collapsed hip" is the hip that is carried lower. However, if collapsed is interpreted to mean a falling together or compacting of parts, then the side of the body where the rib cage is hollow, where the hip bone and shoulder come closer together, would be collapsed. The term "collapsed body" is less misleading.

A collapsed body is most easily viewed from the rear. The pelvis is tilted so that one seat bone is lower than the other. The left seat bone becomes lower, for example, which causes the right leg to hike up and be carried shorter and the left leg to be carried longer. The rider reaches for the right stirrup with the toe, causing the right heel to be elevated. The left side of the rib cage is stretched; the right side of the rib cage is scrunched. To compensate for this crookedness in the lower spine, a rider may adjust her upper body. Often this is done by dropping the right shoulder and raising the left, further stretching the left side and compressing the right. In a final effort to counter-

balance, the head is sometimes tilted to the left to complete yet another S-curve in the spine.

The Back and Upper Body

The rider can use her pelvis, lower back, and weight in a variety of ways, ranging from a soft, light, following contact of the seat with the horse's back to a strong, still stiffening. In most cases, the configuration of the rider's back and pelvis as viewed from the side should be straight, never hollowed, rounded, or slumped. When the back is hollow, the base of the pelvis has rotated backward, tipping the rider forward on her crotch. When the

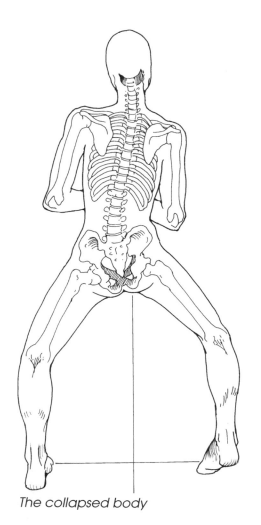

The collapsed body

back is rounded, the rider has let the floor of the pelvis roll forward excessively while at the same time letting the abdominals collapse. This causes the rider to roll back on her pockets. A slumped posture results when the rider lets her entire back collapse and round and does not hold her body up with any degree of muscle tone. Her upper body should be straight without being stiff.

The rider's shoulders should always be parallel with the horse's shoulders: that is, when making a turn to the left, the rider's left shoulder should move slightly back while her right shoulder moves slightly forward. This puts the rider in a position to approximate the arcs of movement of the horse's front legs. The arc of the left leg, in this case, is smaller than that of the right leg.

The Legs

It is important to apply leg aids in the appropriate location on the horse's side and with proper intensity. For most styles of riding it is best to let the thighs relax and allow the lower leg to maintain light, steady contact with the horse's side. If you actively try to use the muscles of your thighs for leg aids, you will probably create muscle tension that will push you up out of the saddle and loosen your seat.

Intensity of the leg aids. In general terms, you use your lower legs actively, passively, or in a yielding fashion. When you want a horse to move, whether it is forward, backward, or sideways, you use an active leg to aid your seat and hands in creating the desired motion. An active leg aid can consist of a squeeze, light bump, kick, or spur.

If you want a horse to continue the motion you have created, you use a passive leg unless the horse becomes sluggish and

This rider has lowered her left hip and collapsed her body on the right side. Note that the right heel is raised and the toe is reaching for the stirrup. The right shoulder is lower than the left. This has caused the horse's hind-quarters to shift to the left.

After adjustment, the rider's body is more square and so is the horse.

needs a reminder or until you wish to change what the horse is doing. A passive leg has contact with the horse's side but is not trying to change things.

A yielding leg is used temporarily when teaching a horse a new movement such as one of the lateral movements. If you want your horse to move away from your right leg and the horse is confused, it might help if you momentarily take your left leg off the horse so as to create a space for him to move into. When you temporarily take your leg off a horse's side, you are using a yielding leg.

Position of the leg aids. Traditionally, the basic riding position calls for the leg to be "at the girth," which is actually a misnomer because the leg in its neutral position is usually behind the girth. The leg in the neutral position initiates and maintains movement in a straight line. As soon as you introduce turning or lateral

On a circle, the rider's inside shoulder comes slightly back and the outside shoulder comes slightly forward. The degree of muscularity of a rider will determine how "square" his or her shoulders will appear.

The first steps of a right turn. The rider's right (inside) leg is in the middle position and gives the horse a reference point around which to bend. The left (outside) leg has been moved back to contain the hind-quarters and keep them from moving to the left.

movements or try to straighten the inherent crookedness of a horse, you will have to begin using your legs in different positions.

When turning a horse, the leg that is inside the arc of the turn often stays in the neutral position to give the horse a pivot or point of reference around which to turn. As you turn to the right, for example, the right leg would serve as this post for the horse to turn around. If when turning to the right your horse arcs evenly from head to tail, then your left leg could also stay in its neutral position. If, however, your horse is like most horses and deviates in the turn or is resistant, you will have to adjust him with various positions of your legs.

For English riding, the lower leg position often said to be "at" the girth would be more accurately described as at the middle position.

The lower leg of a western rider in the middle position.

The lower leg of an English rider behind the middle position is used to activate the hindquarters in a sideways direction.

The lower leg of a western rider behind the middle position.

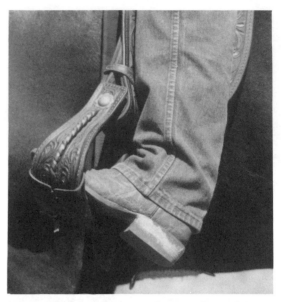

The lower leg of an English rider ahead of the middle position is used to influence the forehand. This lower leg appears loose and unstable, however, as if it might swing forward.

When the lower leg of a western rider is used ahead of the middle position it could cause the rider to brace against the cantle.

If he lazily swings his hips off to the left as he turns to the right, you can move your left leg behind the neutral position to encourage him to carry his hindquarters up underneath himself in the turn. If he is stiff in the shoulders and resists bending to the right as you turn, you can encourage him to pick his shoulders up and move them over by using your left leg in front of the neutral position. And if he resists in an almost disobedient fashion by thrusting his shoulder out to the left, you may have to back up your leg aid with the help of a whip (see the section that follows on artificial aids).

The outside leg is especially important during turning maneuvers such as corners, circles, half-turns and serpentines. If your outside leg comes off your horse's side (which can very easily happen), you lose your balance, the horse loses a point of reference, and his hindquarters will

often drift to the outside of the turn. This will further cause you to lose contact between your outside seat bone and the saddle and will allow the horse to drop his weight on his inside shoulder. Especially after a sharp turn, your outside leg tends to move forward and you will need to expend some energy to keep it back in position to act as a forward driving and holding aid. This is especially critical in a canter, for example, to keep the hindquarter in position in relation to the front end. If your leg consistently comes off your horse, perhaps your stirrups are too long. If you find yourself losing contact with the saddle, perhaps your stirrups are too short, causing you to stand in them.

Correcting your leg position when you are in motion can be tricky. You must learn to move your leg back into position without inadvertently giving other signals to your horse. This is where fitness and the awareness of your body parts as independent entities will really prove their worth. To alter your leg position without shifting your seat, upper body, or hands is difficult, but such intricacies must be mastered or you may imbalance your horse.

Checking your leg position. Generally, if you glance down at your legs (without moving your head or upper body forward) you should not be able to see your foot if your leg is well under your body where it should be. Be sure your ankle is not "broken" inward.

If you ride English, you can examine your boots for an indication of your leg position as you ride as well as other valuable information. First, was it a great relief to pull off your boots? Did their removal restore circulation in your calf muscles or feet? If so, they are simply too tight and probably adversely affect your riding as well as interrupting the circulation in your legs. Consider a pair of custom-made boots or have the tight ones stretched.

Next look at the wrinkles around the ankle of your boots. If you see more than two or three, and they are stacked up on top of each other like an accordion, either the shaft of your boot is too tall and is forced downward into a pile around your ankle, or your boots are too tight at the calf and do not sit in the proper position below the crook of your knee. Compare the number and shape of the wrinkles at the outside and the inside of the ankle. If there is a gross imbalance, it should give you a clue as to which way you break your ankles when you ride.

Now inspect the inside of the shaft for sweat marks. Sweat marks mean you are working your horse, and that is good! But where the sweat marks are located may not be such good news. Do the smudges

Stirrups too long.

Stirrups too short.

Stirrups just right.

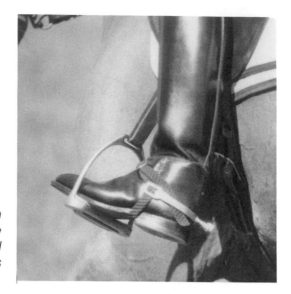

The outside leg in front of the middle position is being used to control a horse's drifting forehand.

The ankle broken inward results in an unstable leg and the lower leg coming off the horse.

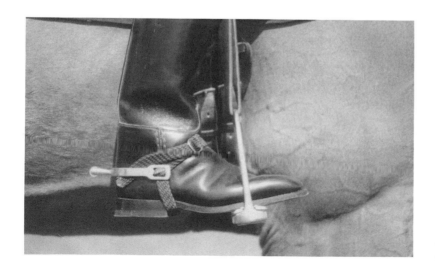

After you ride, "read" your riding boots to see if you held your leg in an ideal position.

and smears run over the mid-seam at the back of the shaft and make it seem more like you are riding with the backs of your legs against your horse? If so, the next time your instructor tells you that your toes point out drastically or that you use the back of your heel rather than your calf for leg cues, you will have a vivid mental picture of what she means.

But sweat marks are the signs only of today's ride. The wear marks on your boots tell the long-term story. Look for the areas on your boots that always seem to require a bit more shoe polish. Start at the top where your boot contacts the stirrup leather and saddle flap. Is the rub mark on the boot dead center on the inside of the shaft? Or does it creep around toward the back seam to keep company with the sweat marks on the calf of your boot?

Look at the foot of your boot. Is there a wear mark on the inside of the vamp at the ball? This tells you that you carry your foot snug up to the inside of the stirrup. Is there excessive wear on the instep or at the front of the heel? This means you often ride with your foot well home in the stirrup. The proper placement of your foot in the stirrup will depend on the style of riding you are practicing.

Compare your right and left boots. Are they symmetric in their wrinkles and creases and rubs and marks? If not, maybe you work in one direction much more than the other—or your horse is extremely one-sided. On the other hand, maybe it is you! The imbalance from left to right may be caused by using stirrup leathers of different lengths, or by an asymmetry in your own body. While your riding boots may not tell you anything new, they may offer bits of information to help you piece together just what your habits are when you are in the saddle and how you might improve them.

The Hands

A horse's mouth is extremely sensitive and any fear of the bit or rigidity in the jaw will be reflected in the whole horse. That's why all rein aids must be applied with smoothness and finesse. Fast jabs just make a horse tense and substantiate his fear of the rider's hands. Uneven reins give incorrect signals. For all styles of riding, there must be some degree of contact.

A too-tight rein never allows a horse to be rewarded, so it is very unlikely that

he will ever relax and move naturally. With all rein aids, a taking of the rein should eventually be followed by a release. If the horse has responded correctly, the rider should release the aid and follow the horse's correct movement. If the horse has responded incorrectly, the rider can either hold the rein until the horse complies or release the aid momentarily, reorganize, and re-apply it. Only in instances of specialized or supervised training should a rider hold a rein against a horse's extreme resistance until he submits to it.

A too-loose rein may not be safe or effective. When rein cues are applied on a too-loose rein, they usually require large movements on the part of the rider as she gathers up the reins. Often such grand motions are big surprises in the horse's mouth. It is better to have moderate contact, no matter what the style of riding, so that you can feel the horse and influence him with subtle movements of your hands.

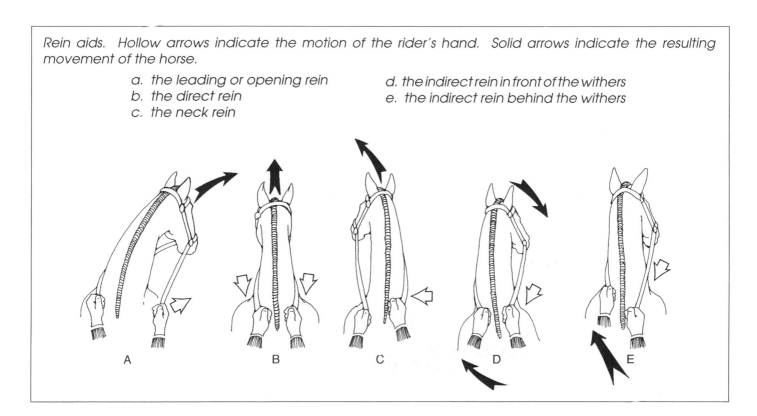

Rein aids. Hollow arrows indicate the motion of the rider's hand. Solid arrows indicate the resulting movement of the horse.

    a. the leading or opening rein   *d. the indirect rein in front of the withers*
    b. the direct rein   *e. the indirect rein behind the withers*
    c. the neck rein

A   B   C   D   E

The aids should be balanced and light.

Be sure the reins are even or you will be giving incorrect signals to your horse.

For rein aids to be most effective, they should be applied so there is a direct line from the rider's elbow through the hand to the horse's mouth.

It is a common error for beginning riders to overemphasize rein aids and it is surprising how many intermediate and advanced riders continue this counterproductive habit. Rein aids should be subtle and should augment or add to the performance that is created by the more powerful aids: the mind, the seat, the upper body, and the legs. The reins should never be used to balance your upper body or stabilize your seat. Your hands must develop a steadiness independent from your other body movements.

Two-handed rein aids. Riding with two hands is common to all forms of English riding and the training stages of Western riding. The various effects of the reins on a snaffle bit riding with two hands are discussed below.

The leading or opening rein: the most elementary rein aid to turn a horse. A pull to the right with the right rein turns the horse to the right by throwing the weight to the horse's right foreleg. The rider's hand should move to the side and slightly forward, never backward. The left hand can give or be passive or supporting. The left rein will limit the degree of bend to the right.

The direct rein: a rein that opposes the front end of the horse to the hind end. It is a pull straight back resulting in a straight line from the horse's mouth toward the rider's hip. A direct rein settles the weight to the shoulder on the same side as the applied rein or toward the hindquarters.

The indirect rein in front of the withers: an indirect rein aid that shifts the horse's weight from one shoulder to the other. The rein moves backward and toward the withers but does not cross over the withers.

The indirect rein behind the withers: an indirect rein aid that shifts the horse's weight from one shoulder to the opposite hind leg. The rein is held behind the withers and moves toward the rider's opposite hip but does not cross the midline of the horse.

One-handed rein aids. When riding western with one hand and a curb bridle, the primary rein aids are the direct rein and the neck rein.

The neck rein: The neck rein is more of a conditioned reflex than an actual rein aid. The horse is taught to turn away from the touch of the rein on his neck. The goal is for the horse to turn left when the right rein touches the right side of his neck. The trainer first teaches the horse using two hands: one hand applies an opening rein and the other applies a neck rein. To teach the horse to turn to the left, a left opening rein is applied slightly before the right neck rein. Gradually, the time between the two rein aids is decreased. Also, the intensity of the left opening rein is diminished until the horse turns left simply from the touch of the right neck rein on his neck. A neck rein is most effective when it is applied just ahead of the withers.

A neck rein must be applied very lightly or it will give the horse the opposite signal than is desired. A strong right neck rein will pull the horse's nose to the right and backward. If you find yourself tempted to use this much rein in an attempt to neck-rein to the left, one or two things need work. Either the horse needs a review lesson or you need to learn how to use your seat and legs more effectively to assist you in turning the horse so you can de-emphasize your rein aids.

Roles of the inside and outside reins. It is customary to designate the reins as an inside and an outside rein. During elementary work, the inside rein is the one

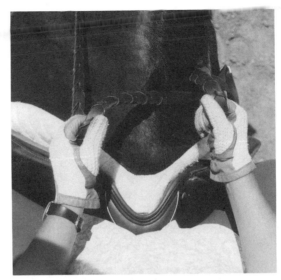

Although the hands are held a bit wide, this is an acceptable hand position.

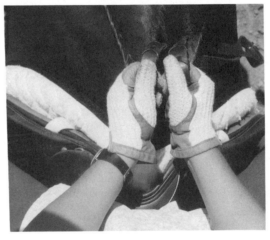

Hands held too close together cause wrists to be broken inward. Sometimes this is called for by an instructor in order to correct a rider's poor habit of wide, loose hand carriage.

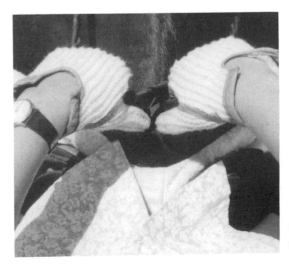

Wrists broken downward, sometimes called "puppy paws."

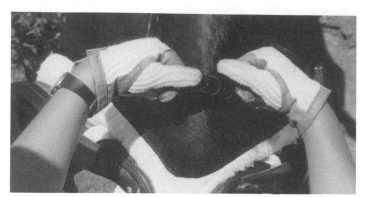

Hands rolled inward. Sometimes effective for a momentary increase in contact but should not be held this way for longer than a second.

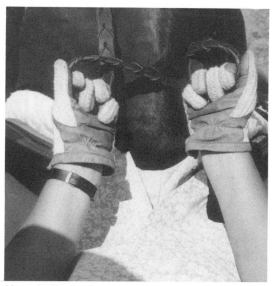

Hands flipped over.

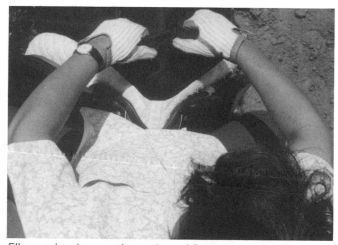

Elbows broken outward and "puppy paws."

nearest the center of the arena, the outside rein is nearest the wall. Correspondingly, the inside rein is the one at the inside of the arc of a turn. However, as maneuvers become more advanced, horse and rider may perform in one direction while bending in another. In such a case, inside would always refer to the inside of a bend, no matter in which direction the horse is traveling in the arena.

The outside rein. In most cases, the outside rein is a stabilizing force. It provides your horse with a constant point of reference as to the gait, speed, and degree of collection in which he should be performing. In general terms, the outside rein should be steady so your horse is allowed to relax mentally. This will encourage him to relax his jaw, poll, and neck into the contact with that rein. If you keep the outside rein steady, the horse will soon realize that the rein is always there and he will conform his mental attitude and the frame of his body to that reference point. If the outside rein is too loose, the horse will likely experiment with falling on his forehand, dropping to the inside, or letting his outside shoulder bulge through the outside rein. If the outside rein is too tight, on the other hand, the horse will likely be tense, either raising his head above the bit or overflexing and dropping behind the bit, or perhaps counterflexing (bending his head in the opposite direction of the turn).

The connection to the outside rein should be from your outside hip bone, through your elbow, wrist and hand, through the rein to the horse's mouth. This boundary that you have created contains your horse's energy and, depending on the degree of your driving aids, gives him a specific parameter for the degree of energy and engagement with which he must perform. And yet this

To develop flexion at the throatlatch and jaw, contact is first established with both direct reins. The hands are held higher and wider than normal for purposes of illustration.

The rider's left hand (inside rein) then moves backward toward her left hip while the right rein (outside rein) maintains its contact and supports the action of the left rein. This causes the horse to relax his lower jaw and flex in the poll.

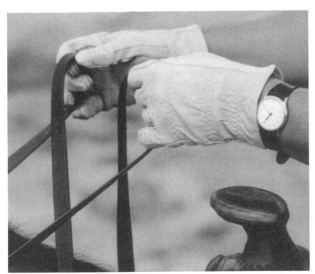

The hands should be carried lightly and in cooperation with each other. These hands are elevated for a momentary check (half-halt).

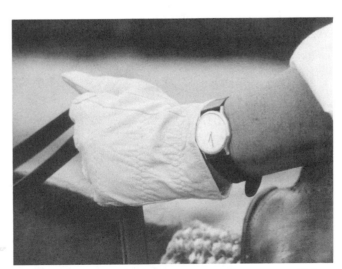

When riding with one hand, it should work within a 4-inch radius of the horse's withers.

boundary is not rigid or unyielding — far from it. It should be a giving hand that yields to the horse in rhythm with his movement. The outside rein should never be "thrown away" during active work or the horse will lose his point of reference for balance and collection. The exception to this is when testing the thoroughness of a horse's training or when allowing a horse a stretch break.

The inside rein. The inside rein can be used steadily or in a subtle, fluid, massaging way to get your horse's attention, to develop flexion, and to shape your horse's form in various maneuvers. It should be used in conjunction with the outside rein to produce harmony. Yet the more a rider learns to use the seat, upper body, and legs, the less pronounced the action of the inside rein becomes.

The inside rein is often slackened for a moment or two when the horse is performing in a desired state of flexion and collection. This is a test of the self-carriage of the horse: Is the horse holding himself in the desirable configuration on his own or is he is dependent on the rider's aids to keep him together? The stronger and more physically developed a horse is, the more he is capable of self-carriage.

ARTIFICIAL AIDS

Artificial aids should augment or reinforce natural aids, never replace them. The spur, for example, can be used humanely to heighten a lazy horse's response to the rider's leg. A typical scenario might go like this:

■ The rider applies the lower leg. The horse does not respond.

■ The rider applies the lower leg, followed immediately by a moderate application of the spur. The horse responds.

■ The rider applies the lower leg. The horse does not respond.

■ The rider applies the lower leg, followed immediately by the spur. The horse responds.

■ The rider applies the lower leg. The horse responds.

A rider who uses the spur for *all* leg cues is losing the effectiveness of a very important tool and may be treating the horse inhumanely. He may eventually become desensitized to the spur and the rider will have to use sharper and sharper spurs with harder jabs to get the horse's attention. If a rider saves the spur for use as reinforcement, however, the horse will remain sensitive to it.

The whip should also be thought of as a reinforcer, not a replacer, of the natural

The spur is a reinforcer, not a replacer of your natural aids. Here the rider has chosen to wear the spur below the spur guard on the boot to put it in a more effective position for collecting the horse.

aids. It can be used in various positions to assist the rider's aids. In some instances the whip can be used to influence a part of the horse's body that cannot easily be reached by the mounted rider's legs. It should never be used as punishment. It should be used with an effective tap, not constant light nagging or an abusive volley of sharp cuts. The whip should be used strongly enough so that the horse wants to avoid it but not so strongly that he becomes afraid. If the horse is afraid, he will not be relaxed and his reflexes will not work in a normal fashion. He will be unable to sort out what he is supposed to be learning.

If a horse is falling in on his inside shoulder or bulging out on his outside shoulder, the rider can use the leg in front of the neutral position on the offending side, in conjunction with the whip at the shoulder, to encourage him to carry himself upright and straight. Sometimes just holding a whip in the strategic position is enough of a reinforcer or reminder to the horse.

If a horse is not bending his body sufficiently for a circular movement, using the whip just behind the girth will assist the inside leg in establishing and maintaining the bend.

A tap with the whip on the horse's abdominal muscles tends to make him contract his abdominals, which in turn

The whip is carried across the thigh to reinforce the lower leg.

brings the hind legs more underneath his barrel. This helps teach the more advanced horse the physical configuration for collection. Similarly, an instructor on the ground can tap the whip on the top of the croup to cause the horse to drop his croup and tuck his tail. This in turn brings the hind legs farther under the body.

Other artificial aids such as draw reins, side reins, and martingales really fall into the realm of the horse trainer so will not be discussed in this book on the training of the rider. If your instructor uses these devices ask for an explanation of their use.

CHAPTER 13

COORDINATING THE AIDS

The more time you spend riding, the better you will be able to coordinate your aids. You will need to identify and use each part of your body independently, yet in cooperation with the rest of your body. There will be times when what your instructor asks you to do seems physically impossible.

Take the aids for the half-pass, for example. The horse is moving forward and sideways, and he is bent in the direction of movement. In a half-pass to the left, the horse's neck is bent to the left and his right legs cross over in front of his left legs. To achieve this, the rider must:

- look forward

- maintain forward driving with the back and seat

- weight the left seat bone to achieve bend and attain direction of movement

- maintain left bend with the left leg in the middle position

- initiate and continue sideways movement with the right leg behind the middle position

- initiate and maintain left bend and flexion with the left rein

- maintain pace and degree of collection and regulate bend with the right rein

Coordination of the aids is possible only if a rider has an independent seat — that is, one that follows the movement of the horse without bracing or gripping. In order to develop an independent seat, it is beneficial to ride bareback or without stirrups. With no stirrups to rely on for stability, you will discover other ways to develop your security: sitting upright and in balance from left to right. Such an awareness will show you how important your seat is for collecting and organizing a horse. You will also realize that when you do ride with stirrups, if you jam your feet forward and brace in the stirrups you tend to pull on the reins as a counter-balancing maneuver. Ride every day without stirrups for twenty minutes or so or until your inner thigh muscles are sore from stretching and reaching. To add to your stability, as you lengthen your legs, focus on stretching your thighs by moving your knees closer to the ground rather than trying just to lower your heels.

Posting. Whether you are an English or a western rider, you will benefit from learning how to post properly. Your instructor may decide it is best for you to learn to post before learning to sit the trot. Posting allows you to ride a horse at a more active trot by rising and falling with the movement of his inside hind leg and the corresponding outside foreleg. So if you are tracking to the left (riding around the ring making left turns), you would rise as the left hind and the right front reach forward, and sit as they land.

The correct diagonal. Your instructor will want you to develop this skill by *feeling* the action of the horse's hind leg to determine whether you are on the correct diagonal or not. As you are learning, however, you may be allowed to check by quickly glancing down at the horse's outside shoulder. You should never lean over to look for diagonals or leads. Use just your eyes; even if you just tilt your head down and forward to look, you will tend to throw your upper body off balance.

If you see that the outside shoulder is reaching forward as you are rising, you are on the correct diagonal. If you are posting on the incorrect diagonal, merely sit two beats and resume posting.

When you rise for posting, you must maintain your balance over the center of your foot. If your foot swings backward or forward as you rise, your stability is greatly reduced. You will grab with your thighs or pull on the reins to help straighten up

To correct the tendency to brace and grip, as illustrated here, the rider must develop an independent seat.

Riding without stirrups allows the rider to discover other ways to develop security and improves the seat and leg.

Posting allows you to ride a more active trot. You must be sure to keep the leg under the center of balance, however, or it may cause you to brace and balance on the reins.

As you are developing a feel for how your horse is moving, never lean over to see what diagonal or lead he is on. Instead, glance down with your eyes only.

and balance. This in turn will make the horse tense and will throw him off balance. Practice standing in the stirrups at a halt with your upper body inclined forward and your weight settled down into your entire foot. Your stirrup leathers should be vertical. Practice maintaining your balance at the walk and trot in this standing position, taking care not to pull on the reins. Resume posting and see if the rising position hasn't become easier.

Sitting the trot. Learning to sit the trot is the key to developing an independent, following seat. You should never attempt to sit still in the saddle at the trot by holding on or bracing yourself. The resulting tension will pop you right out of the saddle. To develop a secure seat at the trot, begin by regulating your breathing and check yourself periodically to see if you continue to breathe rhythmically. Position your pelvis so that your seat bones roll underneath you, not out behind you. Then concentrate your weight over your seat bones by keeping the weight of your upper body directly over the seat bones, shoulders over hips. Keep your back straight but relaxed, and support it with tone from your abdominal muscles. Relax your thighs so that instead of pinching, they open and fall vertically along the horse's sides. Finally, lightly embrace the horse's ribs with your lower legs. If you are riding western or saddle seat, your trainer or instructor might advise you to keep your lower legs off the horse until they are needed.

LONGE LESSON

It will probably take you six months to one year of concentrated effort to develop a good seat or to correct a faulty one. To that end, some instructors choose to work their students on a circle with a longe line. The reason an instructor-controlled longe lesson is so valuable is that the rider does not have to guide the horse and can concentrate totally on her position. She can find her balance on both a stationary and a moving horse; she can concentrate on eliminating imbalances or crooked body positions. Sometimes exaggerations are employed to bring a rider's faults to light and to change a rider's sensory awareness. After experiencing the extremes, a middle ground will be found and the rider will achieve a functional and technically correct seat. The seat must be supple, upright and forward, yet deep and independent from the other aids. Longe lessons also provide a good opportunity for the rider to develop proper leg contact and position, which is directly related to the development of the seat.

The manner in which the school horse works on a longe line will determine the

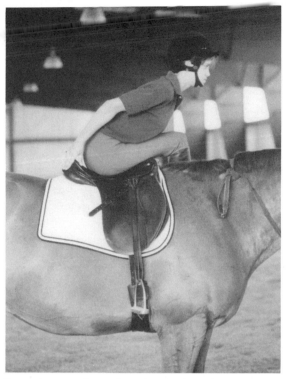

Exercises on horseback can be started at a standstill. To develop a sense of independence, the student swings one leg over the horse's head, then stretches forward.

Your instructor may choose to work you on the longe line to help you develop independent aids.

While keeping the seat and lower leg position steady, the student reaches her arms over her head and rotates from side to side.

Without cueing the horse with her lower leg, the student holds her arms horizontally and swings from side to side.

To develop confidence and balance, the rider brings both legs over the horse's neck and clicks her heels together.

success of the lesson. If the school horse is obedient to the instructor on a 33 foot longe line, then the circle the horse makes is a 66- foot or 20-meter circle, such as is used in lower levels of dressage. If the school horse must be kept close to the instructor, such as on a 10- or 15-foot line, then the small circle will make things more difficult for both the horse and student. The horse will be trotting in a 20- or 30-foot circle, which is more on the order of a volte, a size more suited to collected work. A horse would need to be very strong and obedient to trot in such a small circle in a balanced and correct fashion for very long. A student would also have to work hard on such a small circle and would need to have already developed a fairly good seat in order to benefit from such a longe lesson.

Self-Longe

The self-longe lesson is accomplished without an instructor in a round pen 20 meters in diameter (66 feet across) to approximate the bending required for a 20-meter circle. Footing in the pen should be dry and not excessively deep or hard. The horse is outfitted with a saddle with no stirrups and a snaffle bridle with no regular reins. Elastic side reins are attached to the billets and adjusted evenly, with a contact that is suitable for the horse. The rider should wear a protective helmet.

In a self-longe lesson, the rider will have to initiate gait transitions, but should not have to do more. Therefore, for self-longe lessons, a very steady, well-schooled horse is required. He must be sound and trustworthy. His rhythm must be consistent, with no increase or decrease within or between gaits unless they are called for. He must accept fairly snug contact with the side reins because they will help keep his movement connected.

Lifting the thighs off the horse requires strength and concentration.

To help the student learn the feeling that moving the lower leg off the horse begins up in the hip, the instructor provides resistance at the rider's ankles.

Self-longe lessons can help you develop balance.

He should be physically able and mentally willing to stand quietly at a halt with the side reins attached. A horse that backs, throws his head, or swerves the hindquarters instead of standing at a square halt is not sufficiently on the bit to use in a self-longe lesson.

It might be necessary to longe the horse with a line first to take the edge off his energy before mounting. In any event, for the first self-longe lesson or so, it would be best to have someone on hand to start the horse if necessary and watch to be sure the horse is suitable for such work. It can be very dangerous to mount a horse with side reins attached; it is best to mount and then attach them.

The horse must be relatively balanced from side to side as well as from front to rear so that he performs without falling on the forehand, leaning on the side reins, or drifting in or out of the circle. The self-longe horse must be able and willing to perform the gaits in a quiet rhythm as well as at a more active pace as the rider's development requires. It is not necessary for the self-longe horse to be a spectacular or elegant mover. As long as his rhythm and balance are relatively steady and he works with relaxation, he will allow the rider to work on the same.

USING YOUR MENTAL CHECKLIST

As you are being longed by your instructor or are working in a self-longe lesson, you can begin to go over your personal mental checklist. While a checklist may initially seem to take a long time to run through, it soon becomes a quick series of automatic adjustments. Once mounted, open both legs off the horse, rotate the thigh inward from the hip and place the thigh flat on the saddle. Let the lower leg hang freely for a moment. Sit with your weight settled equally on both seat bones and in the deepest part of your saddle. Use regulated breathing exercises to help you let your weight settle down onto the horse. Now let your lower leg find its natural position close to the contour of the horse's side. Let your foot lie parallel to the horse's side with your ankle relaxed so that your heel is slightly lower than your toe.

Keeping the seat and leg in this good position, stretch first one arm and then the other straight over your head to stretch out your rib cage. Now focus your attention on your upper body. With both arms out to the side, rotate your arms in backward circles to bring your shoulder blades back. Raise your sternum up and forward, further drawing your shoulder blades flat

Mental Checklist for Evaluating Position

Breathing deep and regular
Weight on both seat bones
Thighs relaxed
Lower leg contact appropriate
Back relaxed and following
Trunk straight
Shoulders back
Shoulders level
Head straight
Eyes looking forward
Hands even
Hands proper level
Contact even in reins
Hands steady but following horse
Ear, shoulder, hip, heel alignment
Awareness of any spots of soreness
      or tension

against your back. Keep your head straight, neck back, chin slightly tucked.

Once you are satisfied with your position at the halt, ask the horse to begin walking and note your movements. Are they counteractive and rigid or harmonious and following? On a 20-meter circle, especially at the walk, there is very little need for bending or the lateral aids. You will be sitting quite equally balanced left to right. However, as the gait increases or the circle decreases in size, you will begin to sit more on your inside seat bone. Register this feeling. Also at this time give your shoulders a check. Don't let the outside shoulder slip behind the inside one. Remember, your shoulders should parallel the horse's shoulders, your hips the horse's hips.

When your horse begins trotting, assess whether you have allowed your

body to collapse. Most often, no matter what direction the horse is traveling, a rider will sit with some degree of a hollow right side and a stretched left side, caused by the tendency of most horses to carry their bodies hollow on the right side. (Notice the posture of most people as they drive their cars and you will see that the hollow right side is strongly evident here as well.) This causes the rider to lower the left side of the pelvis, which makes the left seat bone slip off the saddle and the right seat bone rise up. The left side of the rider's rib cage is stretched and the right side is hollow. The rider's left leg is lower than the right leg, which often reaches for the stirrup with the toe.

Correct a collapsed body by making the hollow side (for example, the right side) of your body taller. Do this by taking the opposite shoulder (left) back or even twisting your upper body to the left to bring the floating right seat bone down on the saddle.

To stretch your hollow rib cage, reach your right hand up over your head as you did in the initial positioning sequence at the halt. The other hand can secure your balance, if necessary, by holding on to the pommel or cantle. When using both hands on the saddle, in order to encourage proper shoulder position you should always place the outside hand on the pommel and, if necessary, the inside hand on the cantle. This encourages the outside shoulder to come forward with the horse's outside shoulder on a turn. If you have a body alignment problem, however, you may have to dispense with the ideal recommendation until the problem is corrected. Be sure to have very experienced, qualified help in this area.

Beside constant position checks and subsequent corrections, you can also do various exercises while the horse walks, trots, and canters. Ask your instructor which ones would benefit you specifically or refer to *The Complete Training of the Horse and Rider* by Alois Podhajsky for some suggestions. Simple exercises such as rotating arms, legs, ankles, and head are good for relaxation and preventing stiffness.

Although eventually you can work from thirty to forty-five minutes on a longe lesson, it is best to work up to this amount of time gradually. During your regular riding lessons, your instructor can evaluate the work you have done during your self-longe lessons and offer further suggestions.

TRANSITIONS

A transition is a shifting of gears. Most commonly, an upward transition indicates a change from a standstill or a slower gait to a gait that is more ground-covering. Examples are halt to walk, walk to trot, trot to canter, walk to canter and lope to gallop. A downward transition is a change from a gait that is more ground-covering to a halt or a gait that is less so. Examples are trot to walk, canter to trot, canter to walk, walk to halt. In addition, transitions can indicate a change of movement within a gait. For example, a trot can be regular, collected, or extended, so the change from a regular or collected trot to an extended trot is considered an upward transition.

Good transitions are prompt yet smooth. When you drive your car, do you pop the clutch? Or do you hold it halfway out and let it slide for an extended period of time? Do you push the accelerator to the floor and rev the engine? Or do you barely tap the gas pedal and let the engine die? Is it obvious to someone walking down the street that you have just shifted gears as your car bucks or squeals or sputters? Or do you smoothly coordinate all of the mechanics of shifting your car so that the pedestrian doesn't even know you've changed gears?

When you ride, do you spur and pull on the reins at the same time and wonder why your horse's head goes sky high? Do you let the reins go slack, then spur for a canter and wonder why you're thrown forward and the ride is so bumpy and uncomfortable? Or do you coordinate your aids so that gait transitions are smooth and balanced with very little change in the horse's carriage? Will an observer think that the change came so fluidly it was as if the horse melted into the new gait? Those kinds of transitions are the goal of riding, but it takes practice to develop the precise timing necessary to develop such harmony.

Transitions are a balancing act between your driving aids and your restraining aids. Although the net result of these forces should be near zero, indicating that they approach being equal, there should tend to be more forward motion (drive) in a transition, as if you always have your foot lightly on the gas pedal. This will prevent a horse from stalling, falling apart, or losing his degree of collection during a transition.

TURNING AND BENDING

Because we can't ride a horse in a straight line indefinitely, we must learn how to turn him by bending his body. In fact, circular movements develop a horse's body so that eventually we can ride him in

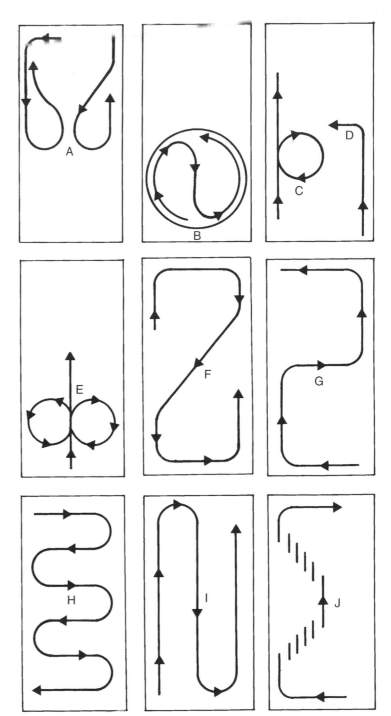

a. half-turn and reverse half-turn
b. change of rein through the circle
c. full circle
d. quarter-turn
e. figure eight
f. change of rein across the diagonal
g. change of rein across the school (or ring or arena)
h. serpentine
i. change of rein down the centerline
j. lateral movement such as leg yield or half-pass.

Instead, from the normal hand position.

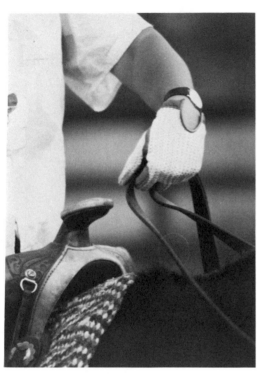

When checking the western horse, follow the sequence of aids in the text. Do not break the wrist as shown and pull back on the reins.

Elevate the hand momentarily.

a truly straight line! Bending can range from an almost imperceptible arcing of the horse's spine on a large circle to an extreme curving of the spine in a sharp turn.

The rider uses her upper body (weight) and seat to tell the horse where he must move. A horse tends to move under and support a rider's weight. So if a rider wants to turn to the right, she must bring her right shoulder back to weight the right seat bone. This will cause the horse to move toward the right. The rider's hands and legs tell the horse what configuration his body should be in, and he moves right. If the turn is to be sharp (such as a quarter-turn), the rider will use a stronger right direct rein and a stronger right leg in the middle position than if the turn were gradual (such as an arcing turn on a large circle). The outside hand and leg (in this example, the left) support the action of the right aids and prevent the horse from bulging out his left shoulder or letting his hindquarters swing to the left. The school figures used in lessons require various degrees and types of bending.

THE PREPARATORY CUES: CHECKING AND THE HALF-HALT

A transition should always be preceded by a preparatory cue or set of aids. An example of this is the preparatory cue you use when longeing your horse. As he works around you on a 30-foot line, you need to be able to get him to change gaits promptly, without catching him by surprise. Many of your longeing signals are a coordination of body language, whip position, and voice. When your horse is trotting and you want him to walk, your preparatory signals consist of lowering your head, not looking at your horse, saying the word "and" with a low-pitched, falling inflection, and beginning to lower your whip toward the ground. Then you deliver the actual cue: you say the word "walk," you hold the whip at its low, walk position, and your horse walks.

Conversely, if the horse is walking and you want him to trot, you might brighten and elevate your body carriage as you look directly at the horse, say "OK" with a chirpy, rising inflection, and begin to raise your whip. Then when you actually give the command "trot" and raise your whip to its trot position, your horse will trot. You have given him time to prepare and you have not surprised him so he is able to do what you ask smoothly.

Similarly, while you are riding, you should prepare your horse for transitions. Of course, you will not be able to use visual cues from his back, nor are voice commands generally acceptable when riding, although they might be used to clarify something for an inexperienced horse. That means you will have to rely on your mind, seat, upper body, legs, and hands to tell your horse that something is coming.

Preparatory aids need to be applied at the optimal time: long enough ahead of the actual cue to allow the horse to ready himself, but not too long before or held for too long or he may get tense or out of position and won't be able to do what you ask. If the horse has responded correctly to the preparatory aid, the aid itself should follow in quick succession.

In western riding, the preparatory aid is called picking the horse up or checking the horse. This is similar to the half-halt in dressage: a calling to attention, a request that the horse reorganize and rebalance

Coordinating the Aids for Selected Maneuvers

Maneuver	Mind	Seat	Upper Body
Upward transition: walk from halt; trot from walk, trot from halt	project forward "There is the spot we are headed to, now let's go; I feel like an arrow."	tighten abdominals to push seat bones forward	stays straight
Downward transition: halt from walk or trot; walk from trot; walk from canter	"I'm going to downshift smoothly so my car goes straight and doesn't lurch."	sit deep letting seat bones move forward, straighten lower back with abdominals, then release	keep shoulders over the hips or very slightly back
Arcing turn to the right using two reins	"Lets bend gradually without falling inside or out."	weight right seat bone without collapsing hip so that hips are in line with horse's hindquarters	right shoulder back, left shoulder forward to stay in line with horse's shoulders
Turn on the haunches to the left, two hands	"Sit down and walk around your hindquarters."	deep and hips even; weight left seat bone to hold left hind leg, let right seat bone move slightly forward in time with the movement of the right hind leg.	keep shoulders back but let right shoulder move forward and over with the movement of the horse's right shoulder.

Coordinating the Aids for Selected Maneuvers (cont.)

Legs	Hands	Additional Comments
close both calves on horse until he responds, then lighten	maintain contact but both hands yield	the same aids are used for all of the above transitions — the difference is a variation in feel, by degrees.
pressure with both calves as horse powers down, then passive contact	firmly fix both hands without pulling back, then yield as horse responds by slowing down	legs should be ready to adjust for straightness of stop. If horse's hindquarters drift to the left, use left leg to control.
right leg at the girth for point of reference for the bend; use left leg at the girth to keep horse moving forward; use left leg behind the girth if hindquarters swing out	right direct rein to the right hip; left supporting rein	between this sweeping, gradual turn and the turn on the haunches (hindquarters), there are many gradations.
left leg slightly behind the girth to hold left side of the hindquarters from moving sideways over to the left; right leg at the girth to initiate and preserve forward movement and the crossing of right front leg over left front leg.	slightly lift up and use right indirect rein in front of the withers to shift the weight to the left hind leg; left direct rein to create a slight left flexion but very minimal bending	the rhythm of the four-beat walk should be maintained.

Coordinating the Aids for Selected Maneuvers (cont.)

Maneuver	Mind	Seat	Upper Body
Turn on the forehand, horse bends left, hindquarters move right, two hands	"Keep your forehand steady and move your hindquarters in a semicircle around it."	not particularly deep; left seat bone slightly weighted and pushed forward to horse's left shoulder.	left shoulder forward to weight the left front leg
Canter or lope, right lead	"Come under behind, come up in front, and roll forward smoothly into a three-beat gait."	right seat bone forward and up; left seat bone back and down	push down on the seat then follow the forward movement (without leaning forward) just as the horse creates the forward movement, not before.
Back or rein back on straight line	"Round up underneath me and move your legs in a one-two time backwards."	even weight on seat bones, flex gluteal muscles to tilt pelvis and bring seat bones forward.	straighten lower back to help seat bones come forward
Half-halt or check	"I want to tell you something. Listen. Get organized. Something is coming. Hello, is anybody home? Attention!"	push seat bones forward, drive, and release	straight or slightly back depending on the forcefulness of the half-halt

Coordinating the Aids for Selected Maneuvers (cont.)

Legs	Hands	Additional Comments
step down slightly in the left stirrup, left leg well behind the girth to move the hindquarters to the right; right leg at the girth to keep horse moving forward (with no steps backward) and to limit the size and frequency of the steps to the right.	left direct rein creates left flexion; right indirect rein in front of the withers shifts the weight to the left shoulder, supports and keeps the right shoulder from bulging out or the right front leg from stepping to the right	horse's nose tips slightly to the left; the left front foot remains relatively stationary.
right leg on girth; left leg behind the girth; both active	right direct rein to create flexion and an appropriate amount of bend; left supporting rein or bearing rein	apply the aids when the left hind leg is about to land
equal pressure with both legs on the girth as long as you want horse to continue backing	lift up and use equal direct rein pressure to initiate movement, then once horse is backing freely, yield rein pressure and continue using seat and legs only	horse moves diagonal pairs of legs; may need to use squeezing or vibration on reins if horse is stuck; never attempt to pull horse backward with reins
both calves and/or spurs on the horse's abdominals at the girth or cinch	gather in an upward fashion, then a progressive, gradual yielding retaining light contact	use before all transitions: as an attention getter; as a means of collecting horse; for elevating and lightening the forehand by engaging the hindquarters and requiring them to carry more weight; to slow down or calm an anxious horse. Rider has a momentary feeling of floating.

his body so that the coming transition will occur correctly and smoothly. For the rider the half-halt is a momentary, symmetric, isometric contraction of the back, the abdominals, the lower calves, and the hands on the reins. The horse should respond in kind, by tightening up his act, so to speak. Remember, this is a momentary aid, meaning it should be followed by a yielding by the rider.

Properly executed, the half-halt or check will make the horse feel like he is in a state of energized suspension for just a fraction of a second underneath you. At that moment he is listening intently to you through his body; that is precisely when you should tell him what it is that he should do next.

Sometimes it takes a series of preparatory commands to get the horse organized enough to make a transition. This would be especially true of a horse cantering briskly and being called down to a walk. The difference in speed may simply be too great to expect the horse to make the downward transition cleanly and fluidly. Using four or five half-halts or checks brings the horse down a notch each time. Properly executed half-halts encourage a horse to develop self-carriage instead of being dependent on the rider's aids holding him together.

For the halt, the rider deepens her seat, balances her weight rearward, keeps her shoulders over her hips, and maintains appropriate rein contact. When the horse responds, the aids are released.

CHAPTER 14

A TYPICAL RIDING SESSION

Atypical riding session includes preparation of the horse, warm-up of horse and rider, the lesson, cool-down, and post-ride care of the horse.

PREPARATION OF THE HORSE

Catching and Haltering

Your relationship with your horse begins with the first step you take toward him to catch him. Whether he is in a box stall or pen, approach your horse with confidence and a specific haltering plan in mind. Don't be threatening or timid in your body language. Always walk toward the horse's shoulder, never his rump or his head. Move at a smooth, slow walk. Don't look directly at the horse, but keep him in your peripheral vision.

Haltering in a safe, organized fashion prevents mishaps and bad habits. When the horse stands and allows you to approach, walk up to him on the near (left) side. Hold the unbuckled halter and lead rope in your left hand. With your right hand, scratch the horse on the withers and then move your right hand across the top of the neck to the right side. Pass the end of the lead rope under the horse's neck to your right hand and make a loop around the horse's neck or throatlatch. The loop is held with your right hand. If the horse tries to move away at this stage, you can effectively pull the horse's head toward you while levering your right elbow into the middle of the horse's neck.

To halter, hand the halter strap with the holes in it under the horse's neck to your right hand, which is holding the loop. With your left hand, position the noseband of the halter on the horse's face and then bring the hands together to buckle the halter.

Hoof Care and Grooming

As you lead your horse to the grooming area, do so as if your instructor or a horse show judge were watching you at all times. Stay in proper position in the vicinity of the horse's shoulder and be direct and precise in your body language.

Tie the horse or attach him to cross-ties in a manner so that he is safe for you to work on while you are grooming and tacking him. If it is fly season, you may need to begin with a light spray of the legs with a fly repellent. This will make it safer for you to work around his legs.

Put on a pair of barn gloves and begin picking out the hooves. For front hooves, stand by the front leg facing the opposite way the horse is facing, and run your hand down the tendons at the back of the cannon. This should signal the horse to lift his leg. If not, pinch the tendon just above the fetlock. Be ready to catch the hoof as the horse lifts it and hold it in one hand while you pick the hoof out with the hoof pick held in your other hand. Teach yourself to be ambidextrous. To pick up a hind leg, stand alongside the hip, facing the opposite

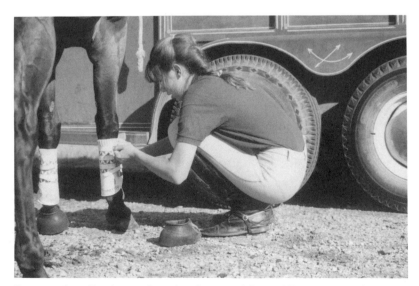

Remember the lower back relaxer while putting on your horse's protective boots.

way the horse is facing, and use your shoulder to lightly push the horse's weight off the hind leg you are near over to the other one. Then run your hand down the leg in a similar fashion and clean the hoof.

Grooming begins by loosening dirt, loose hair, dead skin, and scurf from the horse's skin with a rubber curry on all of the muscled portions of the horse. Use a rubber grooming mitt to perform a similar function on the horse's head and legs. Then with a stiff bristled brush whisk the loosened debris from the body with short flicking motions of your wrist. At this stage of grooming, long sweeping brush strokes only serve to relocate the dirt. Vacuuming is also appropriate at this stage of grooming. Use a soft to medium brush on the head and legs. Once the majority of the dirt is removed, use a soft brush to finish the coat with long, smooth strokes. Set the coat and remove final dust with a cloth, either dry or damp.

Saddling and Bridling

Place the saddle blanket or pad on the horse's back slightly ahead of where it will eventually sit, then slide it into position. Place the saddle on the blanket, peak the blanket in the gullet of the saddle, and secure the girth or cinch. Put on protective leg boots or bandages if necessary.

In order to keep control of the horse while bridling, untie the lead rope, remove the halter from the horse's head, re-fasten it around the horse's neck, drape the lead rope over your left arm, and proceed to bridle.

Hold the bridle in your left hand with the reins draped over your left arm. Reach your right arm over the horse's neck to the off side and bring the right hand near the right ear. Transfer the headstall of the bridle from your left hand to your right hand and then place your left hand on the bridge of the horse's nose. Move your right arm over the top of the horse's neck so your arm reaches between the horse's ears. This brings the headstall in front of the horse's face. Take the bit in your left hand. Move the bit into position between the upper and lower incisors, and then ask the horse to open his mouth by placing your left thumb in the space between the incisors and molars. Once the bit is in his mouth, put the right ear in the headstall, then the left ear. Buckle the throatlatch (and noseband if an English bridle). Remove the halter from the horse's neck and you are ready to lead your horse to the arena.

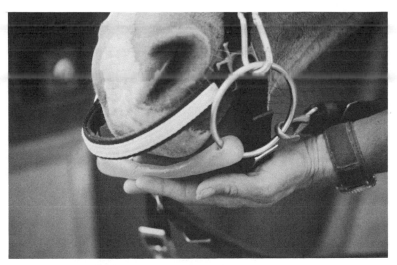

Bridling. This endurance rider chooses a non-metal bit for cold-weather riding.

THE LESSON PLAN

Divide the work session into sections and take breaks between sections. Look at a one-hour work session as a five-part, strategically planned lesson. This will allow you to ride your best and your horse to perform his best. At first you should use a watch to keep track of how long you work in each stage.

Start with the following plan for a one-hour work session and modify it to suit your own situation.

Ten-Minute Warm-Up
Ten-Minute Old Work
(Two-Minute Break)
Fifteen-Minute New Work
(Three-Minute Break)
Ten-Minute Old Work
Ten-Minute Cool-Down

Especially if your daily routine is rather sedentary (school, office) you will need to pay attention to the warm-up to allow you to get the most out of your ride and prevent injury. Sitting causes muscle shortening and body stress. If you don't design your warm-up to prepare your body for effective active movement, your ride may actually add more stress and tension to your life.

The warm-up serves several purposes. It prevents damage from sudden, unusual stresses. It readies the neurological pathways and makes them alert for signals, increasing coordination at the beginning of the work. It pumps more blood to the skeletal muscles, increasing their strength of contraction. It allows the muscles to stretch. First evaluate your mental attitude and make adjustments . . . slow down, relax, breathe. You are going for a ride to enjoy yourself. If you approach your work in a good frame of mind, it is more likely your horse's attitude will reflect your own.

You can actually begin your physical warm-up when you are grooming your horse. The currying provides good circular arm motion and can work to combat sidedness if you groom with both arms and pay attention to how you stand and

Variation of the stirrup stretch.

use the sides of your body.

As you move around to tack up your horse your legs are also getting warmed up. When you lead your horse to the arena, stretch as you walk as if you want to lengthen your torso. Broaden your chest by bringing your shoulders back.

Stretching exercises should not be used as the first part of a warm-up, as they may result in torn fibers. It is most beneficial to get moving first and use stretching exercises later. Your warm-up continues as you lead your horse to the work area.

Leading to arena. Leading a bridled horse is somewhat different than leading a haltered horse. With the latter, a lead rope is attached to a ring under the horse's jaw and you direct him with left, right, and backward movements of the rope. A bridled horse, on the other hand, has a rein attached to each side of the bit, so if you grab onto the reins together and treat them

like a single lead rope, you will give some confusing and contradictory signals to the horse's mouth. So, separate the reins with your index finger and use the reins independently to indicate to your horse whether he should turn right or left or slow down.

When you reach the arena, stop your horse straight and square and give him the command to stand. You should take your time preparing to mount as it will develop patience in your horse. You will need to let down your stirrup irons if you are riding English. Check to see that your saddle is straight, then step in front of your horse and see if your stirrups are even. Give the cinch or girth its final tightening. Put on your riding gloves, sunglasses, secure your hat, and mount.

Mounting. To mount, face the rear of your horse on the near side. Hold the reins, mane, and whip, if used, in your left hand. The reins should be even and there

Lead the bridled horse properly.

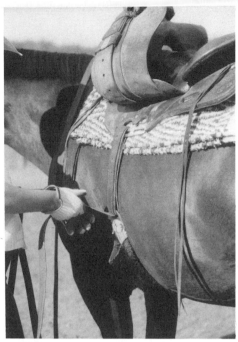

Before mounting, check the cinch or girth.

The rider prepares to mount but her horse is distracted.

should be enough tension on them so that the horse does not walk forward as you mount. If you have too much tension on the reins, however, the horse may become anxious or he may begin to back up.

Use your right hand to turn the left stirrup toward you. Place the ball of your left foot in the stirrup, your left knee against the horse, and your right hand on the pommel, swells, or horn of your saddle. Bounce one to three times on your right leg and on the final bounce rise, thinking "Up and over" as you raise your body straight up and swing your right leg over the horse. Keep your right leg straight and your foot high as you swing your leg over the horse's back so that the knee does not bump into the saddle and you don't kick your horse in the rump. Lower yourself into the saddle by using the muscles of your legs — don't just collapse in the seat.

Sit for just a moment without doing a thing. Then tell your horse "Whoa" as you begin to reach for your right stirrup with

She looks to see if something may cause her horse to spook.

your right foot. You might accidentally kick the horse as you search for your stirrup, so you need to tell him in advance you don't want him to move. When you have both stirrups, gather up and straighten your reins and then sit quietly again for a moment. When you decide it is time to move off, give your horse the appropriate signals and walk on a long rein.

The horse's warm-up. The horse's warm-up should start with a few minutes of lazy walking on a very long rein so that the horse can blow and stretch his back and neck in preparation for more active work. After a few minutes, pick your horse up by sitting deep, flexing your abdominals, putting your lower legs on the horse's sides, and gathering up the reins. For the rest of the ten-minute warm-up, walk or jog your horse either along the arena rail or by making large figures such as 60-foot circles or large serpentines.

Your in-saddle warm-up. Your warm-up in the saddle can consist of a few upper body stretches, arm circles, some leg swings, head rolls, ankle rotations, and leg and arm shakes. Don't waste too much energy in the warm up, however; save it for your lesson.

Pre-Lesson or Pre-Competition Warm-Up

The warm-up strategy that you use immediately prior to a lesson or a competition will be different from the one before your daily ride and should be of your own custom design. Every person's level of fitness and peak of performance are different so you must experiment to find what works best for you. As a base, you can start with the following: One half hour before the performance, begin exercising with enough intensity to induce sweating but not to cause fatigue. Exercise this way for fifteen minutes, alternating hard work (vigorous grooming or posting trot) with stretching (squats to put on horse's boots OR leg stretches to attach spurs). Ten to fifteen minutes before the performance, taper off to a cool-down and officially end the warm-up about five minutes before the performance. Use this time to relax and rest.

THE LESSON

Old Work

Go right from the warm-up into the first "old work" session. Ask your horse to do things that he already does well. If you are an English rider, it may be circles or serpentines at the walk, at the posting trot, or at the canter. This should be active, forward work where you are riding the horse on straight lines, that is, with no lateral work. If you are a western rider, your first old work session could consist of jogging large circles, then small ones, loping the arena at large, and making transitions from jog to lope, jog to walk, jog to halt, halt to back.

These types of maneuvers will slowly begin collecting the horse in preparation for the new work session. Be sure that you have plenty of forward work interspersed with the transitions or the horse may develop a balky or behind-the-bit tendency. Whether you are formally taking a lesson from your instructor or practicing

She assures her horse that everything is OK.

Face the rear of the horse, left hand on reins and withers, right hand presenting stirrup to left foot.

Keeping left knee as close to the horse as possible (this horse is 17 hands tall!) place the left foot in the stirrup.

In one smooth motion, swing up and over, keeping the right leg straight so that your knee doesn't bump the saddle.

at home, every time you ride your horse, it is a lesson for both of you. Whether you approach riding formally or casually, learning is always taking place.

You should concentrate on your seat, leg position, use of upper body, position of shoulders on corners, and consistency and lightness of hands. Identify areas that you need to work on in the new work period.

During the short break, let the reins become gradually longer to encourage your horse to work the bit in his mouth, stretch his neck and back, and blow. Take your feet out of the stirrups and rotate your ankles and legs. Stretch your arms and back. Take care, however, not to "throw the horse away" all of a sudden. A sudden release of the reins will just dump the horse on his forehand. Instead let him gradually stretch down. Feed the reins out to him slowly until he is just moseying around the arena, blowing his nose. As the rest period comes to a close, deepen your seat, add lower leg pressure, and slowly pick the reins up again, gradually adding more contact until you have him working in the state he was in before the break.

New Work

Now, during the new work, introduce the current areas of emphasis in your horse's training, such as lateral work. The new work should be demanding without being debilitating. Since your horse is thoroughly warmed up, he can more correctly perform shoulder-in or two-tracking for you. These exercises are valuable for the dressage horse as well as the reining and western riding horse as they strengthen the horse's hindquarters as well as develop his suppleness so that he can perform fluid lead changes.

During the active work session, first focus on previous work that the horse does well. This Grand Prix horse and rider work on the half-pass.

The new work period should also include more difficult transitions and collected work. The transitions might be canter to halt, walk to lope, etc. If you are an English rider the lateral work may consist of turn on the forehand, leg yielding, shoulder in, half-pass, and so on. The collected work may involve cantering in small circles. If you are a western rider your lateral work might be two-tracking, turn on the hindquarters, and turnarounds. Your collected work might be lope and stop, backing, and loping small circles.

How do you know when it is time to quit the new work? This is an art and science. Usually we (or our instructors) are trying to make a breakthrough of some sort in our riding. If a rider has done a marvelous job but begins to tire and make mistakes after twenty-five minutes, it is

Be sure to give your horse periodic breaks to stretch. Take your feet out of the stirrups and take a break yourself.

time to quit the new work period. If a rider has a mental block and refuses to try what the instructor is asking, the instructor may often work until things become successful even if the rider is physically exhausted. Never quit if you have feelings of giving up; instead, walk on a long rein, rearrange your thoughts, and come back to the work.

If the new work session was effective, your horse will be ready for a slightly longer break. Follow the guidelines of the previous break. Once you have picked your horse back up, you will begin the final work session.

Reviewing Old Work

Don't be tempted to go back to trouble spots that surfaced during the new work. Instead, identify the areas where you may have had problems in the new work and go back to the basics that underlie them. Review and reestablish them in your mind and with your horse so that during tomorrow's ride you will have a better chance of performing the new work correctly. You will find that problems will be diminished tomorrow if you end with a review of old work today. End with something you and your horse do well together. This will reestablish a horse's confidence and interest and will bolster your self-esteem. Old work also gives the horse's mind and body a chance to gradually power-down from the energy level of the more demanding new work.

THE COOL-DOWN

After a vigorous ride, it is important for both your horse's and your own sake, to gradually and systematically wind down from the work. During the cool-down, keep your horse on the bit, but gradually let his frame get longer and more relaxed. The cool-down does not have to consist entirely of walking around on a long rein.

At the end of the work, allow your horse ample time to stretch his neck and back and cool out.

A long posting trot is one of the best ways you can finish up the work as it will encourage your horse to stretch his back and will flush out the lactic acid that has accumulated in the large muscles of his body. Your own cool-down can consist of some arm and leg rolls, stretches, and swinging exercises as were mentioned in the warm up.

Momentarily gather your horse up and ask for a balanced halt. Then use the step-down or slide-down dismounting method. Place your left hand (which is holding the reins) on the mane. Remove your right foot from the stirrup, place your right hand on the pommel or swell and press down with your hand to assist your legs in lifting your seat up and out of the saddle. Keep your upper body straight as you swing your right leg over the horse. Then either:

- face forward, step down behind the left stirrup, and take your left foot out of the stirrup; or

- remove both of your feet from the stirrups at the same time so that when you swing your right leg over the horse's hindquarters, you can slide down the horse's side to the ground.

After a particularly vigorous ride, you may choose to get off, loosen your horse's girth or cinch, and lead your horse for the last five minutes either around the arena or down the road.

If you have been practicing a repetitive exercise in the lesson such as thirty minutes on a 20-meter circle, you should do some stretching exercises for your legs and hips, back, and arms. If you are very hot it is not a good idea to jump into a cold shower, as it will stiffen your muscles. For the same reason, be very careful of spraying your horse with cold water after a hard workout.

POST-RIDE HORSE CARE

Cleaning and Grooming Your Horse

Untack your horse in a quiet and systematic way. Using water to hose down your horse every day is not a good long-term management practice. It results in more problems than benefits. As mentioned, the cold water can actually stiffen your horse's muscles. Also, the daily wet/dry situation can be extremely damaging to the structure of the hooves. Horses' hooves are healthiest when they are kept at a relatively constant dry moisture level. Also, fungus and skin problems can occur when horses are frequently wet and aren't allowed to thoroughly dry.

Untack in a quiet, systematic manner. Let your horse release the bit.

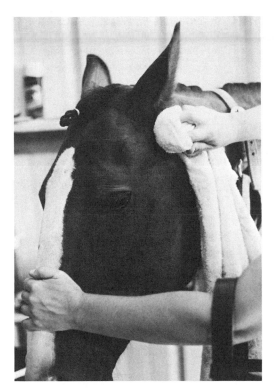

Wipe sweaty areas with a damp cloth.

One solution to cleaning a sweaty horse without hosing him down is to use a body wipe in specific areas such as the head, saddle area, the underside of the neck, and between the hind legs. Body braces are available commercially, or you can make your own by filling a gallon plastic milk container with water, adding 2 tablespoons of Calgon water softener, 2 tablespoons of baby oil, and one ounce of your favorite liniment. This mixture lifts dirt and sweat off the horse's hair, conditions it, and stimulates the skin. If your horse is very sensitive, you may need to decrease or eliminate the liniment from the formula. For any horse, do not use liniment near the eyes, nostrils, or on the anus.

After wiping your horse down, leave him tied so he can dry while you attend to your tack. (Occasionally give your horse a treat by holding him on a long line for five minutes so that he can graze as he dries

off.) Wipe the mouthpiece and rings of the bit with a damp cloth. Spray the leather portions of your tack with a liquid saddle soap. To make your own, fill a spray bottle with one part Murphy's liquid soap to four parts water and use this for your everyday cleaning. Use paste or bar saddle soap for your weekly cleaning. Be careful not to spray the bit with the soapy solution. Wipe all of the sweat and dirt from your tack before returning it to the tack room.

After a sweaty workout, there is nothing a horse likes better than to roll. If you allow your horse to roll in dirt, you will be constantly dealing with dirty saddle blankets and tack, so let your horse roll in sand or shavings. After he has had a chance to perform this act of self-grooming, you can brush him or vacuum him and return him to his stall or pen.

Turning a horse loose follows the same procedure as catching, but in reverse order. First, a loop of the lead rope is applied around the horse's neck, the halter is removed, and then the loop is released. Hold the horse momentarily with the loop,

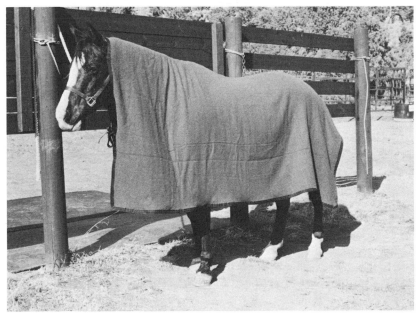

A wool cooler protects this horse from chilling.

Allow your horse to roll. Here the horse is enjoying a wonderful back rub in deep sand.

then release the loop and gently push the horse away with your right arm or hand. If you are dealing with a chronic bolter, you might try dropping a few feed wafers on the ground before you turn the horse loose. He soon will think more about inspecting the ground where he stands than about running away.

After you have returned all of the tack to its proper place, squat in the lower back relaxer position, exhale, and plan for your next ride. Think of one or two things that you did particularly well and one thing that you want to improve tomorrow. Remember that to improve your riding, you have to change. The more open-minded you are to new ideas, the easier it will be for you become the rider you want to be.

Stretch your lower back and plan tomorrow's ride.

PART FOUR

APPENDIX

RESOURCES

Manufacturers of Clothing for Riders

Ariat International
940 Commercial Street
San Carlos, CA 94070
Boots for English and Western riders.

Dover
Box 5837
Holliston, MA 01746
Wide variety of English rider clothing and helmets.

Elser's
P.O. Box T
Hazelton, PA 18201
Wide variety of English rider clothing and helmets.

Frantisi, Inc.
75 Fernstaff Court, #23
Concord, Ontario
Canada L4K 3R4
Boots, breeches, and riding jackets.

W.L. Gore & Associates
P.O. Box 3000
Flagstaff, AZ 86003
Western hatliner.

Miller's
2335 Murray Hill Parkway
East Rutherford, NJ 07073
Wide variety of English rider clothing.

Patagonia
P.O. Box 150
Ventura, CA 93002
A large selection of both summer and winter clothing for active athletes.

Schaefer Outfitter
P.O. Box 4343
Steamboat Springs, CO 80477
Jackets and coats for winter riding.

Miscellaneous Resources

Horsemanship Safety Association
P.O. Drawer 39
Fentress, TX 79622

United States Pony Club
4071 Iron Works Pike
Lexington, KY 40511

4-H Program
c/o your local Cooperative Extension Agent

National 4-H Council
7100 Connecticut Avenue
Chevy Chase, MD 20815

American Riding Instructor
  Certification Program
P.O. Box 282
Alton Bay, NH 03810

American Horse Council
1700 K Street NW, Suite 300
Washington, DC 20006

RECOMMENDED READING

Books

Dunning, Al, *Reining,* Colorado Springs, CO: Western Horseman, 1983.

Feldenkrais, Moshe, *Awareness Through Movement,* New York: Harper, 1972.

German National Equestrian Federation, *Advanced Techniques of Riding,* Middletown, MD: Half Halt Press, 1987.

German National Equestrian Federation Staff, *Principles of Riding,* Middletown, MD: Half Halt Press, 1990.

Hill, Cherry, *The Formative Years, Raising and Training the Young Horse,* Ossining, NY: Breakthrough, 1988.

Hill, Cherry, *From the Center of the Ring, An Inside View of Horse Competitions,* Pownal, VT: Garden Way, 1988.

Hill, Cherry, *Horse For Sale,* New York, NY: Macmillan, 1995.

Hill, Cherry, *Horsekeeping on a Small Acreage,* Pownal, VT: Garden Way, 1990.

Hill, Cherry, *Making Not Breaking,* Ossining, NY: Breakthrough, 1992.

Hill, Cherry, *Maximum Hoof Power,* New York, NY: Macmillan, 1994. (Co-author Richard Klimesh, CJF)

Hill, Cherry, *101 Arena Exercises,* Pownal, VT: Storey, 1995.

Leonard, George, *The Ultimate Athlete,* North Atlantic, 1990.

Loomis, Bob, and Kathy Kadash, *Reining,* California: EquiMedia, 1990.

Morris, George, *Hunter Seat Equitation,* 3rd edition, NY: Doubleday, 1990.

Stashak, Ted S. and Cherry Hill, *Horseowner's Guide to Lameness,* Philadelphia, PA: Williams & Wilkins, 1995.

Swift, Sally, *Centered Riding,* New York: Trafalgar Square, 1985.

Wanless, Mary, *The Natural Rider,* New York: Summit, 1988.

Zi, Nancy, *Art of Breathing,* New York: Bantam, 1986.

Periodicals

California Horse Review, P.O. Box 1238, Rancho Cordova, CA 95741

Chronicle of the Horse, P.O. Box 46, Middleburg, VA 22117

Dressage & CT, 1772 Middlehurst Rd., Cleveland Heights, OH 44118

Dressage Today, 656 Quince Orchard Rd., Gaithersburg, MD 20878

Equus, 656 Quince Orchard Rd., Gaithersburg, MD 20878

Horse & Rider, 12265 West Bayand, Suite 300, Lakewood, CO 80228

Michael Plumb's Horse Journal, Belvoir Publications, P.O. Box 2626, Greenwich, CT 06836

Practical Horseman, P.O. Box 589, Unionville, PA 19375

Western Horseman, P.O. Box 7980, Colorado Springs, CO 80933

GLOSSARY

ADRENAL GLAND a gland located on or near the kidneys that produces epinephrine and norepinephrine, which prepare the body for energy-expending action — "fight or flight."

AEROBIC the metabolism of ordinary and long-duration endurance exercise. Energy is continuously replenished by nutrients and oxygen delivered by the blood to the muscle cells. Carbohydrates and fats are completely metabolized into carbon dioxide, water, and energy.

AGILITY ability to change direction of the body or its parts rapidly.

AGONIST a muscle (or group of muscles) that is the prime mover in an action, contributing to the desired movement by contraction.

AIDS signals from the rider to the horse. The natural aids are the mind, the voice, the seat (weight), upper body, legs, and hands. Artificial aids are extensions or reinforcements of the natural aids and include whips and spurs.

ANAEROBIC the metabolism of short-term strenuous exercise. Nutrients stored in the muscles (primarily glycogen) are incompletely metabolized without oxygen and produce lactic acid, a toxic by-product.

ANAEROBIC THRESHOLD the point (different for everyone) reached after a certain amount of aerobic exercise when the blood can no longer deliver enough oxygen quickly enough, and the body shifts into anaerobic metabolism. After this point you may become short of breath (due to rapid lactic acid build up, oxygen debt, and carbon dioxide dispersion) and will remain so until lactic acid has been cleared away and oxygen has been replenished.

ANTAGONIST a muscle (or group of muscles) that opposes the movement of an agonist or prime mover. Antagonists often provide a desirable resistance and stabilization to a movement.

BACK a two-beat diagonal gait in reverse.

BALANCE ability to keep the center of gravity over the base to maintain equilibrium.

BARS interdental space between incisors and molars where the bit lies in a horse's mouth.

BEHIND THE BIT a means of evasion whereby the horse avoids contact with the bit by over-flexing the neck.

BEND the side-to-side curve of the horse's spine, especially noticeable in the neck.

CANTER the English term for a three-beat gait with right and left leads; has the same footfall pattern as the western lope.

CANTLE the back of the seat of a saddle.

CARDIOPULMONARY that which relates to the heart and lungs.

CHANGE OF DIAGONAL the rider changes the diagonal to which she is posting. See *DIAGONAL*

CHECK the western version of the half-halt.

COLLECTION a gathering together; a state of organized movement; a degree of equilibrium in which the horse's energized response to the aids is characterized by elevated head and neck, rounded back, dropped croup, engaged hindquarters, and flexed abdominals. The horse remains on the bit, is light and mobile, and is ready to respond to the requests of the trainer.

CONDITIONED REFLEX a reaction learned through training.

CONDITIONING the process of developing tolerance to exercise and new capacities of performance.

CONNECTION the relationship between the driving aids, the restraining aids, and the response from the horse.

CONNECTIVE TISSUE body tissues whose primary purpose is to connect body parts: ligaments connect bone to bone; tendons connect muscle to bone.

CONTACT the tightness of the reins related to the level of communication and flow of energy from rider to horse and back to rider.

COOPERATIVE ANTAGONISM the relationship between two muscle groups during a movement.

COORDINATION harmonious working of various muscles in a smooth, correct way with precise timing.

DIAGONAL pair of legs at the trot such as the right front and the left hind; the posting rider sits as the inside hind hits the ground or "rises and falls with the leg on the wall," the outside front leg.

DEHYDRATION the loss of body fluids.

DRESSAGE French for training or schooling; the systematic art of training a horse to perform prescribed movements in a balanced, supple, obedient, and willing manner.

DYSLEXIA incomplete ability to decipher.

ECTOMORPH one who has a slight or slender body build; a body in which the limbs predominate over the trunk.

ELECTROLYTES minerals, able to carry an electrical charge, that are essential in bodily functions.

ENDOMORPH one who has a fat or heavy body build; a body in which the trunk predominates over the limbs.

ENDURANCE ability to resist fatigue and recover quickly. Muscle endurance is the ability to perform many repetitions of the same movement; cardiopulmonary endurance is the ability of the body to continue producing energy.

ENGAGEMENT use of the horse's back and hindquarters to create energy and impulsion to forward movement

EQUESTRIAN of or pertaining to horses or riding; a rider.

EQUESTRIENNE a female rider.

FITNESS the overall state of health and exercise capacity.

FLEXIBILITY the range of motion (contraction and extension) of the muscles and a lengthening and increased resiliency in tendons.

FLEXION characteristic of a supple and collected horse, there are two types of flexion: 1. vertical or longitudinal, often mistakenly associated with "headset." In reality, it is an engagement of the entire body: abdomen, hindquarters, back, neck, and head. 2. lateral side-to-side arcing of the spine characteristic of turning or circular work; often called bending.

FOLLOWING refers to the rider's hand and seat. Rather than assuming a fixed, rigid position, when the horse is performing correctly the rider's hand and seat move elastically in synchrony and harmony with the horse's movements.

FORWARD ENERGY the mental and physical willingness, and in fact, eagerness, to move forward.

GONADAL GLAND the reproductive glands: female ovaries and male testicles; responsible for primary and secondary sexual characteristics.

HALF-HALT a calling to attention, a physical rebalancing of the horse, a prelude to transitions; can be performed at any gait.

IMPULSION the energy and thrust forward from the horse's hindquarters characterized by a forward reaching rather than a backward pushing motion.

INNATE REFLEX a reaction that is present intrinsically; inborn reaction.

INTERVAL TRAINING alternating a period of heavy work with periods of rest or light work; using a target heart rate (rather than all-out fatigue) to signal the end of a work period.

ISOMETRIC CONTRACTION an exercise where muscle tension increases but the muscle does not shorten during the contraction because it does not overcome the resistance of the antagonist.

ISOTONIC CONTRACTION an exercise where muscle fibers shorten, causing movement.

JOG slow western trot.

KINESTHETIC SENSE a sense of awareness (without using the other senses) of body position and action.

LACTIC ACID by-product of anaerobic metabolism that becomes progressively intoxicating and then fatiguing to muscle fibers; can actually stop contraction during strenuous exertion.

LATERAL MOVEMENTS work in which the horse moves with the forehand and haunches on different tracks such as shoulder-in (where there is a bend in the horse's spine yet he is traveling on a straight line) and half-pass (where the horse moves sideways and forward).

LIGAMENTS connective tissue that holds bones together; pliable yet strong, they usually respond to exercise by getting thicker and stronger.

LOPE western version of the three-beat canter: initiating hind leg, a diagonal pair including the leading hind leg, and finally the leading foreleg.

MESOMORPH one who has a medium muscular build; a body in which there is balance between the trunk and the limbs.

NEUROMUSCULAR COORDINATION that which results from nerve impulses reaching the proper muscles with optimal intensity and correct timing.

ON THE BIT a supple and quiet acceptance of the bit characterized by the horse's neck stretching into a contact with the rider's hands through the rein.

PITUITARY GLAND a gland located at the base of the brain important in a wide range of growth and development functions.

RANGE OF MOTION the amount of movement that can occur in a joint, expressed in degrees.

REFLEX an immediate response to a stimulus.

REMODELING the capacity of bones to adapt to new demands.

RHYTHM the sequenced placement of a horse's feet in a gait.

RICE Rest, Ice, Compression, and Elevation. First aid for a sprain.

SELF-CARRIAGE the capacity of a horse to carry itself in balance without being held together by the rider's aids.

SKILL level of neuromuscular coordination.

SPRAIN injury to a ligament when the joint is carried though a range of motion greater than normal but without dislocation or fracture.

STRAIGHTNESS when the horse's hind legs follow the front legs, relatively speaking.

STRAIN injury caused when a muscle or tendon is overused.

STRENGTH greatest force the muscles can produce in a single effort against resistance.

STRETCH REFLEX automatic reflex to contract muscles when they are suddenly stretched.

TENDONS connective tissue holding muscle to bone; usually allow muscles to move bones and respond to exercise by getting thicker and stronger.

THYROID GLAND located on either side of the larynx; greatly affects metabolism.

TRANSITION upward or downward change between gaits, change of speed within gaits, and change of length of movements within gaits.

TROT two-beat diagonal gait.

WALK four-beat flat-footed gait.

INDEX